Other Books by Ben Charles Harris

Eat the Weeds
Better Health with Culinary Herbs
Kitchen Medicines
The Compleat Herbal
Kitchen Tricks

MAKE USE
OF YOUR
GARDEN PLANTS

by Ben Charles Harris

with drawings by Lauren Jarrett

BARRE PUBLISHING/BARRE, MASSACHUSETTS
DISTRIBUTED BY CROWN PUBLISHERS, INC.

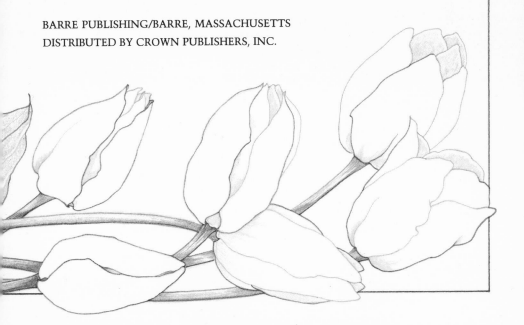

Published simultaneously in Canada by General Publishing Company Limited
First edition
Printed in the United States of America

Designed by Katy Homans

Library of Congress Cataloging in Publication Data
Harris, Ben Charles
  Make use of your garden plants.

  Bibliography: p.
  Includes index.
  1. Plants, Useful — Dictionaries. 2. Plants, Ornamental — Dictionaries. 3. Nature craft — Dictionaries. I. Title.
QK98.4.A1H37   1978   581.6'1   77-17894
ISBN 0-517-53198-4

Author's Note

The information provided in this book is intended for educational purposes only and should not be considered a substitute for the services of a licensed physician. Only he is qualified to diagnose a particular ailment or illness and suggest proper treatment.

CONTENTS

To the memory of my parents,
who encouraged my
further studies in
pharmacology

ACKNOWLEDGMENTS

I AM IMMEASURABLY GRATEFUL to the staff of the Information
Service of the Worcester Public Library and to the librarians of
this city's six colleges. Their relentless and never-failing assis-
tance has made possible the completion of this book.

I would like to express my indebtedness to the listeners of
my People's Health Program (radio) on WICN-FM and the view-
ers of my Green Medicine (TV) show who encouraged me to
prepare a work of this nature.

I owe a debt of gratitude to Lewis Hodgkinson of the Worcester County Extension Service for reviewing the first draft of
the original manuscript, and to the home economists of the
Service for their continual flow of suggestions and for the recipes they provided.

To several members of my herb-study classes, past and
present, I owe special thanks for their testing many of the suggestions contained in this book.

And, finally, I am ever indebted to the legacy of the "old
makers of medicine," to the ancients and early herbalists upon
whose shoulders we moderns stand.

CAVEAT GARDENER!

THE FOLLOWING WELL-KNOWN and commonly cultivated plants
are either potentially harmful or dangerously toxic, albeit
mostly so when taken internally. Take no chances: Omit them
from your future gardening space. In addition, put a DO NOT
TOUCH! sign on the ones marked with an asterisk (*), for these
may cause injury to the skin if you come in contact with them.

Aconite
Anemone
Azalea*
Castor*
Catalpa*
Clematis
Dieffenbachia
Foxglove
Hyacinth
Jasmine, Yellow
Jonquil
Larkspur

Mistletoe
Morning Glory
Mountain Laurel
Oleander
Poppy, Red
Rhododendron
Rose, Christmas
Saffron, Meadow
Snowdrop
Stramonium
Yew*

INTRODUCTION

TWENTY YEARS AGO my neighbor John presented me with all the ripe fruits of his quince tree. "They're no good," he said, "too bitter." A week later I returned and asked him and his sisters for an appraisal of my quince jelly, mixed fruit preserves and sauce, and a hand lotion — all prepared from these fruits. A quick appraisal of my homemade products convinced them to save the fruits themselves and prepare similar products at home.

Sally trimmed her barberry bushes in the fall and had the bagged cuttings taken to the city dump — until she heard my barberry plea over radio station WTAG (August 1954). "Please," I had pleaded with effective theatrics, "please don't discard the wonderful, fall-ripening berries. Use these tart fruits to prepare a jam or jelly or, alone or with cranberries, to yield a zesty sauce." Ever since then, she has prepared similar palate-teasing dishes to her family's great pleasure and amazement. They still find it difficult to believe that the red, football-shaped fruits yield such a delightful sauce for chicken and lamb. And the cuttings, dried and coarsely ground, make a satisfying "health tonic" (herb tea) for Sally.

A dear friend, a flower gardener of the no-herbs-for-me variety, now thinks very highly of his sumac tree, which, for some time, he had wished to eliminate from his property. After a few suggestions from me, he too now realizes that there's more to this unpretentious tree than meets the unsuspecting botanical eye. He has learned to use the early summer dried fruits in a cooling, refreshing lemonade substitute. With the addition of dried orange, lemon, and tangerine peels, he has prepared both a tangy, wholesome marmalade and preserve and, with grapefruit rinds, a "cold-breaker" remedy.

My 1945 talks on "herbs of distinction" over radio station WTAG included aloe, a most popular perennial succulent plant and one of my favorite indoor "utilities." I dwelled mainly on its healing qualities and talked about the fact that its expressed juice was a "jiffy first-aider," etc. A day or two later, a listener phoned to say that she had applied the gelatinous gummy exudation of her "precious ornamental" to a severe sunburn with excellent results; another called to say he had used aloe to soothe insect bites; and a third had found it to be an excellent insect repellent.

Scores of other listeners phoned or wrote to me, relating similar experiences with residents of their gardens: for example, they had used the soft early leaves of the black birch in soups and stews and the summer leaves in wines and beers; the dried, ground meats of the small beechnuts in bread, soups, and casseroles; the *entire* saponaria plant as a shampoo. The stigmas of the crocus (saffron), which are commercially quite costly, were used successfully as a dye by dye expert Mrs. T. J. Healey, and Tom Lee, a student at a local college, told me: "We cooked lots of primrose (and red clover) leaves with meat and fish."

That's what this book is all about.

◆ ◆ ◆

The flora included in this book fall into several categories:

The early birds: crocus, tulip, forsythia, lilac, lady's slipper
The indoor sports: aloe, scented geraniums (pelargoniums), spiderworts, maidenhair (fern), English ivy
The terrarium inhabitants: spleenwort, partridgeberry, sedum
The commoners: marigold, hollyhock, nasturtium, yucca, zinnia, cyclamen
The living fences: arborvitae, privet, barberry, rose
The high shrubs and trees: hibiscus, crab apple, linden, tulip tree
The ground covers: American ivy, bugle, periwinkle
The rock-gardenites: alyssum, columbine, bellflower, pink, bloodroot

The clinging vine: woodbine, Jacob's ladder, mistletoe, virgin's bower
Everyday garden favorites: violet, pansy, saxifrage, hepatica, phlox, lily of the valley
Christmas ornaments: pinecones, magnolia leaves, boxwood clippings

What plants have not been included in this list? For one, plant curiosities like the insect-catching Venus flytrap or the sundew, which are often kept as greenhouse pets and should have no part in a design for a fully functional and efficient garden. In addition, I have not included individual entries on aromatic herbs (although herbs are mentioned throughout as ingredients of recipes and remedies) because their uses are the subject of many other books, including my own *Kitchen Medicines.* I have concentrated specifically on the familiar and popularly grown garden plants, shrubs, and trees commonly found in suburban and rural plots.

Except in passing, I have not discussed the botanical details of these plants — their families, variegated forms, cultivation or propagation, diseases, or techniques of landscaping — because all this information is contained in many other excellent gardening books. Instead, I have focused on the varied, and frequently effortless, uses to which you can put these plants, many of which you may never have thought of. Here are twenty-five practical uses of everyday garden residents:

Foods	*Cosmetics*
Remedies	*Disinfectants*
Tints and dyes	*Hair preparations*
Pillows	*Health teas*
Sachet	*Salt substitutes*
Potpourri	*Deodorants*
Preserves	*Soil enricher*
Jam and jelly	*Insect repellent*
Wine and beer	*Food supplement*
Vinegar	*Tea substitutes*
Seasoning	*Remedies for animals*
Sprouts	*For the bath*
Pickles	

Judge me not too tyrannical a landlord because my garden tenants pay me splendid rent. I am a strict believer in "Ask not what you can do for your garden, but what your garden can do for you." My tenants may be cultivated ones (violet, hollyhock, mallow, nasturtium), transplants from their wild habitat (rose and pyrethrum), or trees and shrubs (pine, spruce, and quince), but for the space they occupy on my land, they remunerate me quite handsomely in dozens of ways. I exile nonuseful ornamentals to a neighbor's or friend's garden, there to join other unproductive species.

Did you know that:

• The fruits of the flowering crab and the hawthorn may be enjoyed in late summer as either a tart nibble, a spiced pickle, or a jam or sauce?

• Yucca flowers may be added successfully to soup or steamed with other foods?

• Marigold flowers colorfully season rice and potato dishes, soups, and chicken broth?

• Nasturtium leaves, a delicious substitute for watercress, add extra zest to salads and vegetable dishes?

• Hollyhock leaves may be enjoyed in gumbo-type soups, stews, and casseroles, and the roots may be used as a remedy for coughs and kidney disorders?

• The dried leaves of the scented geranium serve as a room disinfectant when steeped in hot water, as a flavoring for jam, jelly, and preserves, and as a delicious ingredient in herb drinks?

Planning the Garden

Planning a garden of utility of any size requires a lot of thinking. Bear in mind at all times that what takes up space in your garden must, like the cultivated denizens of your vegetable patch, offer you the greatest returns and benefits for the investment of time and effort. Ask yourself, What are the pros and cons of all prospective newcomers, which are dangerous or poisonous, how much care is involved with each one? In time other questions will pop up: How to preserve and store your newly found friends? Are their fruits or other plant parts worth gathering? What are the problems involved with using them in food and as remedies (and the solutions)?

The gardener should take a cue from the architect or contractor. Neither would dream of building a structure of any kind or size without a well-designed and fully detailed plan. So too should the home gardener design in detail a garden or open grounds so that the plants, shrubs, or trees selected fulfill all necessary requirements: that they be beautiful, suitable for your property, and, most important, useful. Here are some additional hints, gleaned from many years experience in planning gardens for maximum use:

◆ The most common trees and shrubs planted in your yard may serve a useful function as property walls or separators, as well as displaying intriguing silhouettes and showy colors.

◆ Many small-sized trees, such as the crab apple, hawthorn, and dogwood, can yield a wealth of returns on your initial expenditure; their flowers are not only highly decorative, but their tart fruits await plucking from early summer to late fall.

◆ Shade-loving shrubs and (small) trees like the oak, maple, fragrant sumac, and willow can become your protective shield from the sun's hot rays in summer and act as windbreaking shelters from cold gales in winter. This quartet is well suited for slightly acid soil. Their only need: a moderately heavy blanket of fall leaves on the ground over their immediate growing area.

◆ Even a combination of dampish soil and sparse shade should not discourage you in your horticultural endeavors. This combination calls for the following plants: forget-me-not, bee balm, pansy, Joe-Pye weed, cranesbill, fragrant thimbleberry, jack-in-the-pulpit, lady's slipper, day lily, and spicebush.

◆ Make use of usually neglected growing spaces and plant such useful plants as mint, periwinkle, sedum, and violet, for example. Not only will they serve you well, but you'll have efficiently increased the output of other available space with minimum effort and practiced the traditional art of intensive gardening.

◆ Ground covers effectively cover the ground in place of grass and ornament shady places. They need a minimum of care and yield a maximum of value. Thyme, violet, periwinkle, lily of the valley, bugle, and American woodbine are a few worthy examples. They are low, creeping plants whose dense and ever-increasing growth can become rich sources for herbal teas, food seasoners, fumigators, and disinfectants, and medicinal remedies.

◆ Be an adventuresome homeowner and take advantage of *foundation planting.* Plant your selections close to the actual stone foundation of your house or to the adjacent borders and walks. A lilac shrub serves well in an open corner; yew, boxwood, and privet along borders and as guards to the entrance of your home; and spireas conceal any possible dullness or unattractiveness of the foundation's masonry.

◆ For your once strictly ornamental rock garden, try these beautiful and useful plants: alyssum, pinks, sedums, bloodroot, campanulas and the aromatic mints, thyme, pennyroyal, and marjoram.

◆ For your window box, patio, or roof garden: chives, parsley, sage, scented geraniums, lemon balm, marjoram, savory, thyme, mints, fennel, and basil. These plants are also well suited to the apartment dweller because they may be grown in flower pots. (Consider yourself lucky if your back or side yard can afford a sunny 4-by-10-foot space for an herb garden. That's enough room to grow all the food seasoning herbs you will need for an entire twelve months.)

◆ Indoor gardening is a must during fall and winter (at least in the upper regions of the United States). I have confiscated most of the sun-filled windowsills in my home — in the kitchen, utility room, living room, dining room, and three bedrooms — for plants, as my long-suffering wife of thirty-nine years can testify, but at least these almost unlimited spaces have provided us with a continual source of nasturtium leaves, onion and garlic greens, and other fresh edibles and food seasoners for soups, salads, casseroles, and other dishes. To emphasize: Try to make all available space in your home produce as many useful "plants of distinction" as possible.

◆ That little extra space against the garage can be transformed into a border of several beds of sorrel, dill, parsley, chervil, and chives, interspersed with marigolds. And to remedy a sunny but alopecic garden spot, try a ground covering of bugle, mint, thyme, or chamomile as a live soil "toupee."

Harvesting and Drying of the Plants

It is best to gather the plants, or their parts, on the *second successive sunshiny day:* and in the case of aromatic ones, in the forenoon, when the fragrant oil is at its optimum.

Use a sharp knife or scissors.

Cut the plants and gather the flowers just before blossoming. Removing only the upper half of plants provides for new growth.

A general rule for gathering edible greens (hollyhock, nasturtium, et al.): Collect them from spring to early summer. Save for future use by quick-freezing them (and fruits) in cellophane bags. Seal them and label the contents.

The object of drying the plants is to preserve their usable parts and prevent their decay.

There are several ways to dry your gathered material:

1. Tie bundles of the cut plants with string and suspend them, stems up, as close as possible to the ceiling of a well-ventilated room —*away from the sunlight.*

2. Use the attic. Spread them on paper over a clean floor.

3. Suspend the plant bundles near or over, *not on,* your oil burner. There, too, dry the loose plants, seeds, and roots on window screens, or trays of fine mesh wire, or porous cloth.

To prevent molding, do not crowd or overlap the herbs.

Shake or turn the material every third day.

Tag all the materials with their names. Don't depend solely on your memory.

Preservation: When leaves are dry, grind them up coarsely,

having previously removed the stems. Preserve the material in soap-washed and thoroughly dry glass jars or metal or plastic canisters. Label each with the common and Latin names of the plant and date of collection.

Store any excess in a moisture-free area — the attic or cellar.

◆ ◆ ◆

Take a gentle hint from the Shakers' maxim, "The truly useful is always the truly beautiful." Then using this book as a Baedeker, go on to enjoy a more personal and profit-sharing relationship with every tree, shrub, and plant (as well as weed) on your property — and your neighbors' and your cousins' and your aunts' . . .

Many of my students and listeners/viewers of my radio and television shows tell me that they have learned a new appreciation of everything in nature. No longer do they discard the ultranutritious weeds of their flower and vegetable gardens or tolerate flowering do-nothing plants, but are inspired to create new uses for everything from the garden. They are finding their new hobby exhilarating, creative, and fun, as I hope you will.

THE PLANT DICTIONARY

A

Ageratum
A. houstonianum

When I visited central Mexico in 1972, I discovered that the (entire) *febrifugal* (literally, fever-chasing) variety of plant is still much employed as a folk remedy for curing feverish colds and occasional fevers. In similar fashion, the leaves and flowers of dwarf blue ageratums can be used. Gather them from plants bordering a garden, walk, or rockery. Steep* a heaping teaspoonful of equal portions of the dried and finely ground parts† of an aromatic such as mint or marjoram and enjoy a cupful of sickness prevention.

Alkanet
Alkanna tinctoria
Dyer's Alkanet, Bugloss, Redroot

If your hobby is dyeing with various herbs, vegetables, or other plant life, then you'll love this blue purplish flowered perennial. Call it dyer's alkanet, bugloss, or redroot (the Greeks called it *anchousa*, paint, in reference to the plant's dyeing property), and you'll remember to use it to color wines, vinegars, cotton, and silk.

The dyestuff consists of the late-summer roots of the plant, dried and cut into inch segments. Add enough to the wine and vinegar to color and keep the container on or next to a warm

---

*Unless otherwise directed, to "steep," or "prepare an infusion," is to stir a teaspoonful of ground parts in a cup of hot water, cover for 10 minutes, and strain. Sip the liquid slowly.

†Unless otherwise indicated, all parts are dried and finely to coarsely ground.

stove, oil burner, or other source of heat. Shake the contents daily.

To make a cloth dye, boil a tablespoon of the roots in 1½ quarts of hot water down to half, or until the desired shade is obtained. Strain, add the material, and simmer for a few more minutes.

Perhaps your hobby is woodworking and you'd rather prepare your own rosewood or mahogany stain. To obtain a suitable red coloring, immerse the roots in enough commercial benzene to cover for a week or so, remove the roots, and let the liquid evaporate. The yield is a thick paste that's readily soluble in alcohol, (vegetable) oil, and colorless polishing wax. When using either the oil or wax, be sure to rub it into the wood well.

Be prepared for unexpected fresh wounds, insect bites, sores, or minor skin irritations with a jiffy remedy. Prepare your own by simmering a teaspoonful of the dried, cut roots in a cupful of vinegar or sesame or peanut oil for 15 minutes. Or soak the roots in alcohol for a week or so, shaking the container every day. Strain. Wet a cotton ball or cloth and press onto the affected area.

An Interesting Anecdote

During the early 1940s, I phoned Mr. Joseph Buyniski, a fellow druggist, to have him act as interpreter for a Polish lady who had asked me for what sounded like "elchan." She was sure, she insisted, that I had the herb in stock or she wouldn't have driven for almost an hour and through five towns and . . . Yes, said Mr. "B," she wanted alkanna and would use the root (and whole herb) as her family and friends had done "in the old country" for liver and kidney problems.

Aloe
Aloe vera, A. perryi

Grandma Snyder phoned me one afternoon to ask me specifically what to apply to her forearm, which she had just

Aloe

scalded with a potful of boiling water. And her herbal-pharmacist son-in-law suggested: "Boil tea leaves until black and allow the liquid to cool. Meanwhile pluck 2 or 3 of your aloe's largest and thickest leaves. Slice them lengthwise, press out the gummy juice, and spread it evenly over the burn. Let it dry. Alternate each aloe application with a compress of the *cold tea decoction*." Two days later, Grandma exclaimed: "It's a miracle. Healing very nicely." And a big thank you.

Mesdames Lillian Ford and Althea Kately, former members of my herb-study class, related identical wart-removing experiences with aloe. A small piece of expressed juice was well rubbed into the wart every 2 hours or so and more added before covering with a Band-Aid. Every other day castor oil was similarly applied. After 2 or 3 weeks of such administration, the wart had shriveled to a harmless nothing and had almost completely disappeared.

While shopping in a Miami Beach supermarket several years ago, I observed several people buying very large aloe leaves — at least 2 feet long. Two or three shoppers told me how they were going to use them — "best thing for sunburn . . . to prevent sun-dried skin . . . for insect bites . . ."

To obtain aloe's active healing benefits, use a plant over 3 years old. Cut off a large leaf, as close to the stem as possible, wash it clean, and use a knife to remove the sharp spines from the leaf's margin. Cut open the leaf and with a rolling pin (or reasonable facsimile) or a knife's dull edge express the juice into a receptacle. Preserve the exudation by allowing it to dry in an open dish, or by very gently heating it over a warm stove, radiator, or oil burner (the latter's my pet spot). The resultant solid, opaque mass may be broken up into small pieces and stored in an airtight container.

You may use the juice in several ways: apply it on hands or face, as a "spot remover" and improver of the skin, as a burn/ scald remedy, and as a remedy for all skin affections, eczema, insect bites, poison ivy lesions, fresh cuts and abrasions, and ulcerations. (Its antibacterial property is well documented.) Cut a leaf open (lengthwise) and spread it out. Gently pat the juicy inner part upon the affected area and let dry. Do this several times.

To a small open wound or slowly healing ulcer, place a thin transversely cut slice of the leaf and cover with a bandage or Band-Aid. The latter will stay moist and will not stick to the sore spot; the skin will heal, become smooth and pliable, leaving no dense white scar.

A Hair Dressing
After you've washed, rinsed, and toweled your hair (but not to dryness), rub or comb small amounts of the gel thoroughly into the hair and scalp, and brush vigorously.

A Remedy for Crow's-feet
First warm these eye-corner wrinkles with warm water and dry quickly. Apply a thin spread of the gel and massage in gently. Allow to dry and repeat twice. Do this 3 times a day. Or you may use a mixture of equal parts of the aloe and lanolin cream, and apply similarly.

A Better Skin Cream or Lotion
Scrape out the inner gummy liquid of a 10-inch length leaf, express the juice, and incorporate into a tablespoon of your favorite hand cream or 2 teaspoons of lanolin cream or lotion. Apply as needed. Both preparations will help to soften and moisturize the skin better.

My Famous Tea-Aloe Burn Remedy

Boil 3 or 4 (used) tea bags in 2 cups of hot water (covered) for about 10 minutes. Stir in a heaping teaspoon of Irish moss, cover, and let cool. Strain and add 2 teaspoons of fresh aloe juice (or ½ teaspoon of the dried juice). Pour into a bottle and refrigerate when not using.

Permanent Production

Because of their alcohol or ointment bases, the following preparations are stable and need not be refrigerated. They are good to keep on hand in your kitchen for dabbing on fresh cuts and minor wounds.

1. To 4 tablespoons or 2 ounces of rubbing alcohol (ethyl 70 percent), add a teaspoon of the dried gum (or a heaping tablespoon of the fresh exudate). Shake well, strain, and store.

2. To 2 ounces of the above, add a teaspoon of fresh spruce or pine pitch and a teaspoon of dried marigold flowers (rays). Let macerate for 3 or 4 days, shake daily, and strain. These solutions are for fresh cuts and minor wounds, so label the ingredients.

3. Allow formula #1 to evaporate to a thick mass, add enough vegetable oil to smooth out and incorporate into unsalted lard, Vaseline, or lanolin ointment.

See also Pine, Quince, Spruce.

Alyssum
A. saxatile, Lobularia maritima
Snowdrift, Sweet Alyssum

Feeling upset, angry, or nervous? You may infuse a teaspoon of the dried, ground leaves and tops of this old-time plant, once called, appropriately, madwort, in a cup of hot water and drink the strained infusion several times a day. This tea was used in Grandfather's time to calm the nerves and overcome anger, especially after a quarrel between lovers or members of the family. Centuries before, however, the Greeks knew of it: their word *lyssa* means rage or rabies, thus madwort.

Enjoy alyssum's sweet-scented, brightly colored flowers. They do well as a low border edging or ground cover, and once you've adapted them to your window and porch boxes and hanging baskets, you'll have an ever-present supply of fresh herbs at arm's reach.

Amaranth
Amaranthus caudatus
Tassel Flower, Love-Lies-Bleeding

This variety of amaranth invites you to take a culinary cue from the native cooks of southern India and Burma: tint your carrot, celery, and other vegetable soups, or your jellies and jams a beet shade by adding a pinch of the *dried* flowers. (See Crocus and Marigold.) Oriental peoples use the seeds as a cereal and the leaves as a cooked vegetable. Worth a try!

Amaranth, from the Greek word meaning "unfading" or "unwithering," signifies the herb's unfading dark red flowers. Thus, the drooping and slender spikes of flowers are often used in flower arrangements. Also on this account, poets made the plant an emblem of immortality, and the Greeks used it as a decoration for the images of their gods and idols.

In August, after the beautiful clusters of purple red flowers of your prince's-feather or velvetleaf (*A. hybridus hypochondriacus*) have fully blossomed, cut the plants in half and dry by suspending them, bottoms up. Twenty years ago, an English gentleman told me that his grandparents had used the whole herb in treating urinary problems. As an herbal pharmacist, I too have found the leaves and tops of the common amaranth efficacious as a soothing diuretic.

Amaranths are best noted for their astringent-healing property. Especially good to heal ulcerated or bleeding gums and as an effective gargle for a sore throat. Boil a tablespoonful of the late-summer leaves in 1½ pints of hot water for 15 to 20 minutes, let cool, and gargle with this soothing liquid every hour or two.

Anytime you're gardening and accidentally scratch or cut yourself, or are stung by a bee or mosquito, bite a large amaranth leaf — no, don't swallow the juice! — apply it to the affected spot and keep it in place with adhesive or a Band-Aid.

Aquilegia
A. vulgaris
Columbine

This hardy and handsome perennial, well known as columbine, truly deserves your attention. The entire plant is useful in healing all parts of the body.

Gargle
In 2½ cups of hot water, simmer for 10 minutes a tablespoon of the leaves or 1 teaspoon of the fall-collected root, 1 teaspoon of marigolds, and ½ teaspoon each of sage and cinnamon pieces, a remedy for a sore throat or mouth. And if you'll sim-

Aquilegia

mer or gently boil the glutinous (mucilaginous) flowers in the resultant liquid, you may use the strained solution to heal scratches, sores, and other skin troubles.

When you're temporarily discomforted by a minor dysfunction of the kidneys or liver, drink a tea of the leaves, flowers, and seeds (1 teaspoon to a cup of hot water) every 2 hours, with fresh fruit as your only daily intake.

Kidney-Liver Remedy
Mix equal parts of the leaves, dandelion leaves, alkanet, hepatica, verbena, and nasturtium. Steep a heaping teaspoon, finely ground, in a cup of hot water. Drink a cupful 3 or 4 times a day.

Arborvitae
Thuja occidentalis
American White Cedar

Charlie put up a screen fence of arborvitae shrubs around his yard, and when he trimmed them, he dumped the cuttings onto a nearby empty lot. The day we met there, he was unloading them. I asked him to transfer two of his seven bags of cuttings to my car. This he did. But his genuine surprise and sharply raised eyebrows clearly said, "What kind of a nut is this?" I dried the trimmings in the cellar and attic, amidst my wife's cries of anguish. ("Oh, no! You've made a mess along the hallways and up the stairways and *all* over.")

All you'll need for the following kitchen-made products is about two large handfuls of the leaves, cloth material, and a few everyday staples. The shrub's strongly balsamic and camphoraceous aroma invites you to substitute the leaves for the moth-chasing products that contain camphor or paradichlorobenzene. Your attic will benefit as mine has for nearly 28 years by doing the following: Suspend a loose cluster of 6-inch lengths of leafy twigs in each window so that the warming sunlight will strike it well and at length. The pungent vapors will soon disseminate throughout the attic area. (And hook clusters of yellow-buttoned tansy and arborvitae to the attic's ceiling and space them every 4 or 5 feet.)

Prepare several 4-by-4-inch antimoth sachets thus: Stuff cloth bags with a mixture of 5 parts of the dried, coarsely ground arborvitae leaves and 1 part each of rosemary, tansy flowers, and wormwood or southernwood, and other available aromatics, and a few cloves. Seal the bags and place them in clothes or linen closets, in drawers, among or between summerized or winterized clothing, and especially among attic-stored clothing.

Your dog or cat now enters the scene. Prepare a flea remedy for the animal. Use a spice, coffee, or meat grinder to pulverize the above sachet ingredients plus 2 parts each of fennel seeds and pyrethrum flowers. Sift the mixture. Rub the resultant powder all over the animal, well into the hair, and as close to the skin as possible. Add a little more to the neckline. Strew the remaining coarse leftovers over the animal's abode, kennel, or box.

Need to quickly disinfect or deodorize a room, attic, or cellar? Copy the "incensier" (herb-steaming) method used in many French hospitals to remove odors from sickrooms and prevent the spreading of contagious infections. Mix equal parts of lavender, rosemary, arborvitae, tansy, mints, pine (leaves), and thyme, plus a few juniper berries, and stir a cupful in 1½ quarts of hot water. Place the container on a warm radiator or small heating apparatus. Very soon you'll note that the sweet-smelling steaming vapors have extended throughout the room and have cleansed the area of any stale mustiness or undesirable odors.

Here's a no-cost remedy for that painful bruise, sprain, strain, or charley horse. Place a few sprigs of arborvitae, tansy, wormwood (or sage), and any of the other herbs mentioned above in only enough hot water to cover. Stir and cover for 5 minutes. Wrap the warm-hot herbage in muslin or cloth and apply it to the affected areas. Wet another cloth with the remaining liquid, place it on the poultice, and on that keep a hot hot-water bottle. (Though I've sold many commercially prepared products in my pharmacy, I prefer this old-fashioned procedure. A little bother but it works quite well.)

A rub compound for the same problem: Place ½ cupful of the coarsely ground ingredients listed above in a widemouthed bottle and add 2 cups of equal parts of cider vinegar, rubbing alcohol, and spirits of turpentine. Let this digest on a warm radiator or oil burner for 10 to 15 days. Shake, strain, and label.

Or you could prepare a counterirritant for muscle strains and bruises. *Gently* simmer a heaping tablespoon of the ground plants in about a cupful of melted unsalted lard for 15 to 20 minutes. Strain and let it congeal before using. It's a good healing ointment too.

Aster
Aster

Did you know that you can eat the early larger-sized leaves of these strikingly colored plants? That they may be cooked in soup or stew, with fish, and with vegetables? That asters are related to lettuce, dandelion, sunflower, and Jerusalem artichokes? That a warm tea of the leaves is deemed an effective antispasmodic and astringent in diarrhea and stomach complaints?

There are two general kinds:

a. Chinese aster, *Callistephus chinensis,* an annual.

b. The hardy asters or Michelmas daisies. This is the true aster genus.

The Chinese variety was discovered by a Jesuit missionary in the early 1700s. This delightful annual appears in many forms and colors that are long-lasting, a good reason for its being a well-favored resident of the home garden. Here, as Thomas Green points out, they're "particularly adapted to adorn large borders."

The true aster is a hardy strikingly colored perennial and is related to other useful plants such as zinnia, marigold, cosmos, and coreopsis and to such foods as lettuce, endive, and sunflower. The latter association should indicate to you that the lower, early larger-sized leaves of the asters are edible.

Their mildly balsamic taste is a minor challenge but they can become more palatable when cooked with starchy vegetables like squash and zucchini. Take a cue from the American Indians who cooked and ate the spring leaves with fish. Also include them, fresh and chopped, in bean, lentil, or vegetable soup, stew, or coarsely ground, in your fish or chicken patties.

The American Indians utilized native asters in other ways: as an ingredient of a summertime tea, which served to quench thirst or cure a headache. (A decoction of the leaves was applied to the forehead as a compress.) An infusion of the leaves and flowering tops was deemed an effective, carminative, antispasmodic, and intestinal astringent.

Here's an effective and easily prepared antispasmodic/astringent remedy for diarrhea and stomach complaints. Mix 3 parts aster leaves with one part each of bee balm and hibiscus or hollyhock leaves and stir a heaping teaspoonful of this mixture in a cup of hot water. Keep covered for 10 minutes. Stir, strain, and *sip slowly* one such cupful every hour as needed.

The Potawatomi used the flowers of the aster to drive away evil "spirits" working against the patient's recovery; and they used the root as a diaphoretic, aromatic, and antirheumatic.

The Chippewa knew various species for diverse purposes. A cooked decoction of the *A. nemoralis* root was poured into a diseased ear with a spoon. The fine tendrils of the root were smoked with tobacco by hunters to attract game.

From the stems of an unidentified species, states the *U.S. Dispensatory*, the Pawnee Indians prepared moxa: "These are small combustible masses used to produce an eschar (or cautery) by being burned in contact with the skin."

Most Interesting

An aqueous extract of the leaf of the common aster is effective *in vitro* against bacteria such as *Staphylococcus aureus* and *Escherichia coli*. Note that extracts of the New England variety of aster, *A. novae-angliae*, are reported to inhibit only the former; but the whole flowers were once worn as charms.

B

Bachelor's Button
Centaurea cyanus
Cornflower

Warning! Do not cultivate your colorful bachelor's button flowers close to your corn patch or you too will find out why they've earned the names of cornflower and hurt sickle (in Britain):

> Thou blunt'st the very reaper's sickle and so
> In life and death becom'st the farmer's foe.
>   (Author anonymous)*

Permit none to wander from the prescribed bed; otherwise you can expect other unwelcome visitors — troublesome weeds — to invade and steal the sorely needed nutrients intended for your corn and other garden produce.

Late summer asks you to gather your share of the striking blue flowers. Dry them on paper or cardboard here, there, and everywhere. And put them to good purpose.

Are your eyes sore, inflamed, or irritated? Merely steep a pinch of the flowers and chamomile in a cup of hot water, cover, and allow to cool. Stir and strain carefully through filter paper or absorbent cotton. Drop into the eyes several times a day.

For a gentle pick-me-up or appetite restorer, infuse a teaspoon each of barberry and bachelor's button leaves in a cup of hot water and cover until cool. Strain and drink one such cupful 3 to 4 times a day. Or you might try steeping your Pekoe tea with the flowers and leaves.

The herbalist-doctors of the 1800s and today's herbalist-naturopaths suggest that the freshly gathered large leaves of bachelor's button, crushed and slightly dampened, will heal recent wounds or bruises and pus-filled bleeding sores.

---

*Maude Grieve, *A Modern Herbal*, vol. 1. New York: Harcourt, Brace & Co., 1931.

It was shortly before World War II and I was filling prescriptions in my herbal pharmacy. "Do you have cornflowers, Joe-Pye weed, and chicory?" The questioner was a ruddy-faced, tall, white-haired gentleman. (I called him "Mr. Sparkling Eyes" ever after.) He was a Rhode Island herbalist and revealed that he'd used the blue purple flowered plants for folks with blue or purplish faces. Cyanotic conditions? Poor or little blood circulation? Yes, and for liver and gallbladder problems.

I also recall that Mrs. T. J. Healey, the self-taught dye expert of my herb-study class, obtained several bagfuls of bachelor's button bright blue flowers from member flower-gardeners. The expressed juice of the florets, a well-recognized but hard-to-find blue coloring matter, would yield, with the addition of alum, a lasting transparent blue close to ultramarine.

Stella B. Forrest, 93-year-young poet, organist, lecturer, and oldest member of the classes, used the same bachelor's button blue to good effect in her watercolor paintings.

Balsam
Impatiens balsamina
Touch-Me-Not

It's nice to have this pretty annual reside in your garden and display its colorful blooms in late summer. No wonder a synonym for it is touch-me-not and its generic name, *Impatiens.* Touch the ripe pods and the seeds will suddenly scatter.

It's much nicer, however, to have balsams pay rent for the space they occupy (a major theme of this work). For example, you may want to dye white pieces of cotton or wool a light orange or deep yellow. First prepare your coloring base by slowly boiling 5 or 6 large handfuls of the crushed, freshly gathered plant, flowers and all, in a gallon of hot water for one hour. Then strain, reduce the heat, and simmer 4 ounces of the material in the decoction for 45 minutes. Get the shade by throwing a few rusty nails into the boiling solution. The result ranges from light orange to a full red.

Balsam

Here's a quick and effective remedy for bee and insect bites and fresh nettle rash. Crush a few stems of balsam, express the juice, and apply it to the troubled spot. For temporary relief of poison ivy, apply the easily extracted juice every half hour. Meanwhile, boil several handfuls of cut, crushed balsams in 1½ quarts of hot water down to half. (Better if you can add sweet fern, obtainable at your herbalist or health-food store.) The decoction contains a large content of magnesium sulfate, the druggist's Epsom salts, which, when dissolved in hot water, soon alleviates the pain and discomforts of sprains and bruises.

Barberry
Berberis

Granted, barberry bushes make a darn good protective hedge around the garden and lawn to keep out marauding kids and wandering dogs, but the shrub's various parts can be used to even better advantage. From mid-October to December (in New England), there's a plentiful supply of red football-shaped fruits, which can serve us in many ways — barberry can be converted into a tart jam, jelly syrup preserve, or meat sauce, a thirst-quenching drink, and several homemade medicinal remedies.

Syrup
Use 2 pints of fruits to 1 pint water. Barely cover the fruits with water and let soak for 10 to 15 minutes. Cook in a covered pot* for 30 minutes. Strain and add 1½ cups of raw brown sugar or honey. Boil until very thick and bottle. Label and keep in a cold place or refrigerate.

Pickled Berries
Dissolve ½ cup of raw brown sugar or ¾ cup of honey in a pint of cider or wine vinegar and add a cup of the fresh fruits, the *dried* rinds of an orange, and a thin sprinkling of anise and fen-

*Use stainless steel or glass or other nonaluminum or noncopper ware.

Barberry

nel (or other herbs/spices of your choice). Cover and boil only until a definite red appears; and when cold, strain and bottle. Use the liquid in a sauce or relish.

Wined Berries
Gently warm (2 or 3 minutes) white cider or malt wine in a covered pot. Use enough liquid to cover ripe fruits. Sweeten with sugar or honey.

Jelly
Grind a cupful of fresh berries in a blender, remove, and boil with a cupful of sugar to a good consistency. Strain. Let cool and refrigerate. Use with meat, fowl, and fish.

Dried fruits may be first minced in a coffee or spice grinder, barely covered with water, and processed as mentioned.

Don't waste the leaves. Use them fresh or dried. They're a good replacement for lemon. Cook them with meats, fish, chicken, etc. Include them in meatballs and fish balls or patties, casseroles, in basting/marinating mixtures.

The leaves may be used to flavor Pekoe tea. Stir ½ teaspoon of whole or ground leaves in a cupful of the prepared tea. Cover for 5 to 6 minutes, and enjoy.

Thirst Quencher
Cook 2 teaspoons of the dried† fruits or a scant tablespoon of the dried leaves for 5 minutes in a cupful of water. Express the juice and mix with a cupful (or 2) of sweetened‡ water. If over-tart, decrease the amount of the berries and/or leaves or increase the water content. This makes a refreshing cooling drink on a hot summer day. Syruped, a good remedy for throat irritation or soreness.

---

†To dry all parts for future use, suspend entire barberry cuttings in the attic, in the cellar, or over/near a warm oil burner or kitchen stove; or let them dry on the attic wooden floor or on an old window screen. Try not to crowd the material. Shake or turn them every other day. Store each part separately and label the date of collection.

‡Raw brown sugar, honey, or maple syrup. Please avoid white sugar.

Barberry "Lemonade"
Use 2 teaspoons of the ground twigs (cuttings) *and* ½ cupful of the dried berries to a pint of hot water. Boil 15 minutes or steep for 1 hour, add a teaspoon of mint and/or hollyhock leaves, and cover until cool. Stir and strain. This is a refreshing drink for feverish colds and a gentle but active remedy for dyspepsia and stomach problems. It's especially good for liver and gallbladder problems, hence jaundice berry. Dilute ½ cupful equally with water and sip slowly 3 times a day, morning, midafternoon, and evening. Use this remedy as a mild astringent for cankers, aphthae, or sore mouth.

An Amazing Fact
Berberine, the plant's active bitter principle, states C. F. Leyel, England's foremost herb authority, "is more closely allied to human bile than any other substance."

Sauce or Relish
Use it as a tart and flavorful replacement for cranberries, lemons (and other citrus fruits), gooseberries, and raspberries. Use any of the above recipes as a starting point.

Skin Lotion
Boil a tablespoon of the dried and coarsely ground stems (and other parts)§ in 2 pints of boiling water down to 1 pint. Allow to cool, stir, and strain. Pat on every hour as needed for skin problems.

Eye Lotion
Steep 1 or 2 ground dried fruits and ½ teaspoon of chamomile flowers in hot water for 5 minutes, until the slightest yellow appears. Strain through several folds of cheesecloth or cleansing tissue; or use filter paper. Allow to cool and use as drops or lotion every 3 to 4 hours. Refrigerate the rest. Warm slightly when needed.

Dye
To obtain a fast, beautiful yellow dye, boil any and all dried

§You may also include hawkweed, hound's-tongue, and water lily (roots).

parts—fruits, stems, or roots in hot water. The roots yield the strongest shade, especially to wool, and boiled with the usually discarded stems and fruits present a splendid yellow to linen (and leather).

Hair Rinse

Blondes will have even greater fun and enjoyment if, after shampooing their hair, they will rinse their hair in a barberry decoction. Boil a handful of uncut discards or a level tablespoon of mixture (leaves, stems, root) in 1½ pints of hot water for ten minutes. (For extra benefit, stir a heaping teaspoonful of dried marigold flowers into the decoction, after the cooking stops.) Next strain; let the solution cool until it is lukewarm. Now rinse the hair for a few minutes and vigorously massage the liquid into the scalp. Keep hair in a towel-turban until nearly dry and then brush the hair with firm strokes. The barberry rinse helps to accentuate the original color and brighten a once-dull shade.

Bee Balm
Monarda didyma, M. fistulosa
Wild Bergamot, Oswego Tea, Red Balm

The American colonists needn't have worried about having no tea to drink after the Boston Tea Party, for all along they had an excellent substitute for China tea in bee balm. In fact, the residents of Oswego, New York, used so much of it that its popularity quickly spread to the neighboring New England states and thereafter it was called "Oswego tea."

Call it what you will—even the floriculturist's *Monarda*—but gather much of the flowering herb, dry it by suspending it, stems up, and crush or grind it coarsely. It's far more refreshing and aromatic than ordinary tea, and it contains none of the harmful caffeine of tea and coffee, which is now forbidden by physicians to patients with kidney or heart problems.

Whether you're an inveterate drinker of Pekoe tea or not, you'll enjoy that beverage more (with none of its possible side effects) if you stir ½ to 1 teaspoon of the pleasant-tasting balm leaves in a cup of prepared tea. Cover 10 minutes and strain. And to enjoy that cupful as did our grandparents, sip it slowly.

The aroma closely resembles that of bergamot orange (orange mint, *Mentha citrata*), which is a good reason to consider the leaves and flowers when preparing jellies or jams. Try sprinkling a few chopped leaves and flowers over a vegetable salad. Serve them with poultry, meat, and fish. Use the leaves as an ingredient in marinades for meat or venison. And, in general, let them soak in wine or herb vinegar a few minutes before adding them to cooked meat or fish or other foods.

Nervous? Overtired? Slightly nauseated? Stomach distressed? Drink a tepid tea of the leaves and flowers (a heaping teaspoon to a cup) every hour or two. For that canker mouth sore or throat irritation, use a *warm* infusion to rinse or gargle as often as needed. The plant owes its healing efficacy and powerful antiseptic qualities to its rich source of natural thymol, a strong and safe germicide.

The plant is also called bee balm simply because bees love its blossoms, which secrete profuse amounts of nectar. John Gerard, an Elizabethan herbalist, suggested that beekeepers rub their hives with the aromatic herb since "it causeth the bees to keepe together and causeth others to come unto them." For that fresh beesting, rub on some of the expressed juice.

Feeling ill? Drink a warm tea of the dried leaves, and repeat, if necessary, every 2 or 3 hours. Feeling well? Do the same, especially as a Pekoe tea replacement. One of my health-preserving "secret formulas": Using equal parts of bee balm, lemon peel, peppermint, and linden, steep a teaspoonful in a cup of hot water for 15 to 20 minutes, stir, and strain. Enjoy, enjoy.*

The infusion is pleasing and aromatic and provides a safe, effective febrifuge and diaphoretic for fevers and colds, and a stomachic and antispasmodic in cases of colic pain, liver and gallbladder disorders.

*In all recipes and remedies, you may use any other mint: spearmint, apple, curly, round-leaved, bergamot, etc.

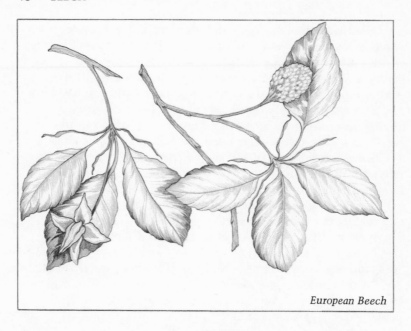

European Beech

American Beech, European Beech
Fagus grandifolia, F. sylvatica

In autumn, gather the small brown twin nuts of the beech tree and set them aside for 1 or 2 weeks. During the "rest" period, they tend to sweeten and become quite tasty. Their highly nutritious meats boast 42 percent fat, 22 percent protein, very high amounts of potassium and calcium, and substantial amounts of phosphorus, chlorine, silicon, and sulfur.

Include the nutmeats in soups, casseroles, stews, and other cooked preparations. Dried and coarsely ground or powdered, they make a good ingredient in bread, muffins, and pancakes.

Grandpa's friends, the Indians, taught him to find large stores of beechnuts in logs or hollow trees, where they were placed by various food-saving rodents. He dried the fall twigs and bark and cooked them in hot syrup to make a serviceable expectorant for bronchial disorders. He also simmered the vegetal parts in vinegar, lard, suet, or chicken fat to create a valuable healing and antiseptic remedy for various skin diseases.

Begonia
Begonia

Did you know that you too may adopt a custom once widely accepted in the East Indies and Burma and use these leaves (of the *Barbata* variety) as a nutritious food? The washed leaves, their *tengoor*, are chopped and prepared, says authority E. L. Sturtevant,* as a potherb for soup or as a cooked vegetable. The Sikkimese of northeastern India use the cooked stems not only as a pleasantly acidic foodstuff; they prepare a sauce of them as do we with rhubarb stems.

Bellflower
Campanula

Enjoy your erect, stoutly stalked purple bellflowers, *C. rapunculoides.* But know that you too may use their creeping root stalks and sweetish cordlike branches to good advantage. Quickly brush-scrub them with cold water and include them in a vegetable salad or with cooked (steamed) foods. You'll like the somewhat parsniplike flavor.

It's like having your cake and eating it too. The *Rapunculus* species is the most plebeian of all and is grown as an ornamental plant in European kitchen gardens because of its graceful, blue violet flowers *and* its roots and young shoots. Europeans eat the former uncooked or, with the latter, steamed. The warmed shoots are pleasantly sweet yet veil a slight pungency. Use the leaves similarly, raw in salad or cooked.

*Notes on Edible Plants, edited by U. T. Hedrick. Albany: J. B. Lyon Company, 1919.

Black Birch
Betula lenta
Sweet Birch, Cherry Birch

Birches can grace your property, lending their beauty to a drab area or to the side of a rocky ridge or to an unoccupied fence corner. They'll also serve you as worthy windbreakers. Spaced in a row, several feet apart, they guard against the onslaught of a northeaster or a winter blizzard.

If yours is an artistic talent, strip part of the thin, smooth bark beneath the outer one, flatten it out with a few heavy books, and use it as a background for a small painting or picture. Perhaps you'd care to use its paper-thinness for a proverb, Christmas greeting, or birthday wish.

In springtime the birch's sweet sap flows. Puncture the tree's bark and the saccharine substance oozes out. Use it as is or to replace — very sparingly — honey, maple syrup, or sugar. If you're a lil' ol' winemaker, add the leafy twigs and the sap to the other ingredients. Allowed to ferment even longer, the wine will become very strong.

Use the spring to early summer leaves too, either fresh or dried in soups, stews, casseroles, and with steaming vegetables. Before you prepare turkey stuffing, meat or fish balls, chowder, or a cheese omelet, soak the tender leaves (minus their stems) in a little basil, garlic, or vinegar for about 15 minutes.

Historical Fact
In 1861, during the Civil War, following their defeat at Carrick's Ford and as they retreated to Monterey, Virginia, hundreds of Confederate soldiers probably saved their lives by eating the edible bark of the black birch, which is full of nutritious sap. "For a number of years after that," states one unknown authority, "the route the soldiers took could be traced by the peeled birch trees."

When you're temporarily plagued by a festering sore, eczemalike eruptions, or other minor skin conditions, apply either a. a lotion or b. an ointment prepared with the tree's fall-gathered twigs, bark, and wood.

a. In a covered pot, boil vigorously 3 handfuls of the parts in 1½ quarts of hot water for 45 minutes. When cool, stir and strain. Apply wet soaks, or use as a compress every hour. Make fresh each day.

    b. Grind the dried parts as finely as possible. Slowly simmer half a cupful in a cup of unsalted lard or equal parts of lard and lanolin for a half hour, occasionally stirring the mixture. Strain and allow to congeal. Apply every 2 to 3 hours.

False Bittersweet
Celastrus scandens

  "Oh, oh. Here they come again," my neighbor would say each late fall, as I and my sons, Irwin and Herbert, approached their rear yards. Armed with our favorite cutters, we'd clip enough of the colorful bittersweet twigs, with their yellow orange seed capsules, to stuff my car's trunk and backseat area to full ca-

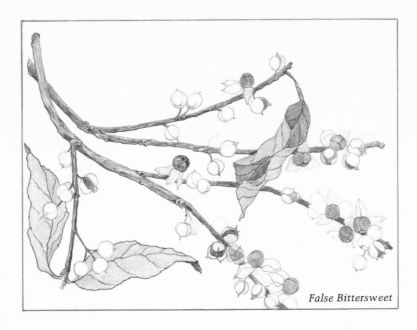

False Bittersweet

pacity. These we'd process and sell retail in my pharmacy and wholesale to florists. This sort of enterprise — combined with their work on our herb garden, drying and cutting the gathered herbs and weedy nondescripts, and working in the drugstore — became the means by which all four of my sons earned their college tuition.

Grandpa had never read the *U.S. Dispensatory*, as I had, to learn that the dried stems of bittersweet, which he (and I have) substituted for sarsaparilla, were there stated to be an internal cleanser or alterative. Experience was his teacher. Invariably, during March and April, our hot kitchen stove would be adorned with a large, hot potful of his favorite bitters or spring tonic, containing dandelion, barberry, bittersweet, etc.

Make your own spring health tonic by simmering a small handful of any part of this trio, together with such aromatics as marjoram, mint, or thyme, in 2 quarts of hot water for 20 to 30 minutes. Let it cool, strain, and undersweeten with honey or raw sugar. Take a half cupful several times a day. The decoction, or a reasonable facsimile, has been found to heal acne, skin eruptions, and boils, and serves as a satisfying corrective in disorders of the intestinal and urinary systems.

The climbing vine has recently gained great prominence as a possible antibiotic and thus a specific medication in the treatment of blood and skin problems.

Bloodroot
Sanguinaria canadensis

Long before my older two sons earned their Boy Scout merit badges for their wide knowledge of edible and remedial herbs, I had stained their homemade bows and arrows with bloodroot juice, just as my grandfather had done for me. Bows and arrows are perhaps passé these days, but this handsome spring flower has many other uses.

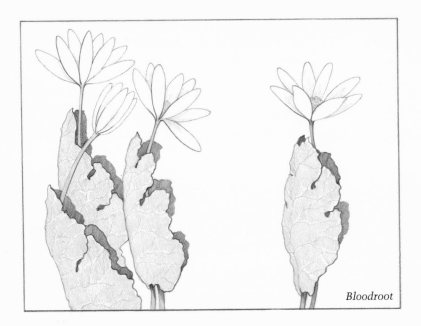

Bloodroot

Enjoy discoveries on your own and have fun experimenting with native vegetable dyes. To produce a dye ranging from orange to red, gather the roots in spring when the plants are in full bloom. To dye wool, cotton, and linen, you'll need 2 to 3 ounces of the dried roots to each gallon of water. The range of yellow orange to red dye will depend on how long the liquid is boiled and which mordants (alum, tin, or chrome) help best to fix the color.

You may also use rubbing alcohol to dissolve out the root's color; and if you dip paper or cloth in that solution, the color becomes a salmon pink. Nor can you entirely wash out the color from flannel or silk stained with the juice of bloodroot.

You may use either of these two remedies to heal a fungus, an indolent open ulcer, or slowly healing sore. Either simmer 2 teaspoons of the dried, ground roots in a pint of hot water down to about half the content; or simmer very gently 3 or 4 teaspoons of the prepared roots in 2 ounces of melted lanolin, mutton suet, or unsalted lard for 20 minutes and strain.

Box
Buxus sempervirens

This evergreen grows naturally to a height of 12 to 15 feet but is usually dwarfed to shrub size (2 to 3 feet) and much used as a garden hedge or border around one's property. Mrs. M. Grieve, England's renowned herbalist, says that some writers have suggested that a decoction of the ground late-autumn leaves and twigs (add a heaping cupful to a quart of boiling water) of the box tree, rubbed in vigorously after a shampoo, will promote hair growth. It may also turn the hair slightly auburn. In either case, it's worth a try.

Box

Bugle
Ajuga reptans

This creeping ground cover is a perennial worth knowing. Once you recognize that its generic name *Ajuga* may well be a corruption of the Latin, *abigo,* to dispel (disease), then you too will use to economical advantage the overground portion of this plant in late May and June when it's in flower. Dry the leaves and flowers and grind them coarsely.

Cut yourself while gardening? Wash a leaf clean, chew it or cut its veins, and press it onto the cut. A bleeding sore or external ulcer, you'll find, will respond speedily to its healing astringency. For that purpose, boil a handful of the washed leaves in a pint of hot water for 15 minutes in a covered pot, strain, and allow to cool. Apply as a wet soak every hour, several times a day. For bleeding or internal hemorroids, dilute the strained warm solution with tepid water and inject every 3 or 4 hours.

To ease an irritating cough, simmer a teaspoon each of bugle's dried ground leaves, hollyhock, and thyme leaves in a pint of hot water for ½ hour in a covered pot. Let cool and strain. Add enough raw brown sugar or honey to produce a syrup. Sip a tablespoonful *slowly* every 2 or 3 hours as needed.

If yours is a gastrointestinal, liver, or gallbladder problem, make a tisane (tea) of the above three ingredients, a good substitute for the now physician-forbidden and irritating tea or coffee. Steep a teaspoon in a cup of hot water for 10 minutes, strain, and add ½ teaspoon of honey — a good health drink.

C

Cardinal Flower
Lobelia cardinalis
Red Lobelia

A discussion of the therapeutic benefits of lobelias (the blue is best known as asthma or emetic weed) must involve Grandpa's simple method of preparing his "asthma medicine." To adopt his procedure, simply place a small handful of the coarsely ground plant, including the seedpods, in a widemouthed jar, add enough of a warmed aromatic vinegar (mint, basil, or garlic) to cover, secure the bottle cap, and let stand 2 weeks. Shake the bottle once daily. Strain, and add honey or raw sugar to thicken. Sip teaspoonful doses every 2 to 3 hours to better produce the desired effect.

For a sore or scratchy throat, stir a teaspoon of *fresh* lemon juice in ½ cupful of the above liquid, and *slowly* sip a teaspoon every hour or two.

For the past 45 years, I've had an ample supply of long-spiked cardinals, those colorful perennials adorning the do-nothing flower gardens of friends. And so, during one winter, when Dr. Franz Meyersohn, newly arrived from Germany, and old-time practitioner Dr. Lewis E. Bishop, both of Worcester, Massachusetts, requested that *fresh* lobelia, i.e., collected in the previous late summer, be used in their herbal prescription, they expected (and received) moderate success from my expertise (ahem!). Their unorthodox lobelia-containing medicines proved to be a safe and soothing relaxant and expectorant in asthmatic and bronchial complaints, and helped control associated spasms.

Carnation
Dianthus caryophyllus

Let's look for a moment at the generic name for carnation. *Dianthus* is from the Greek *dios*, "divine," and *anthus*, "flower," because of its exquisite beauty and fragrance. Hence in Greco-Roman times, it was called "flower of the gods" and used to aromatize wine (and possibly foods). *Caryophyllus* means cloverlike scent and taste, which may help you understand how this flower can be used successfully in the kitchen.

Add the fresh flowers to a vegetable salad, soup, stew, casserole, or pudding. Brave soul, before you incorporate them in an omelet, chop them fine and add them near the end of the preparation. You may use them, fresh or dried, as an ingredient in bread, cake, other baked goods, and pancakes.

Let a conserve of the flowers provide your base for a fowl or fish relish. Prepare a syrup of raw sugar or honey, add a tablespoon of cider or herb-flavored vinegar, stir in 2 tablespoons of fresh flowers, and simmer very gently for 7 or 8 minutes. Cover until cool. The conserve may be taken as a cordial or, with more carnation blossoms, transferred to white wines (e.g., Chablis, sherry, Sauterne), this to become the base for salad dressings, sauces, or pickling liquid for assorted vegetables, Jerusalem artichokes, and nutmeats.

Perhaps, you'd rather aromatize your own homemade wines and beers with "sops in wine," a meaningful synonym for carnation. Use the proportion of a cupful of the near-dried flowers to a quart of slightly warmed wine, shake well, cover, and set aside for 10 to 15 days. Shake the bottle daily. Strain and label the ingredients. Preserve the leftover flowers and include them in a salad or with cooked vegetables. (See Violet.) An excellent vinegar results if this herb wine is allowed to go sour. (*Vinegar* from the French literally means "sour wine.")

The beermaker adds a cupful of the flowers for each gallon of expected return.

For optimum internal therapeutics, select only the petals of fully bloomed flowers, those with the deepest red hue and strongest aroma. As a mild but efficacious remedy for temporary indigestion or other stomach distress, mild headache, or nervous complaint, drink a tea of the dried petals (a teaspoon to a cup of hot water) every 2 or 3 hours and eat mostly fruits during the day.

Taken warm, the infusion, simmered 3 to 5 minutes, becomes an excellent remedy for feverish colds and urinary problems, without the least irritation.

Chrysanthemum
Chrysanthemum

Don't just use mums as a table decoration. If the Chinese philosopher Confucius and his compatriots could enjoy this edible for almost two thousand years, so can you. In fact, the Japanese became so enamored of its values that it inspired a national holiday, the Feast of the Chrysanthemum, celebrated in October. At the end of the eighteenth century, as a result of East-West trade, it became a familiar sight in England and finally emigrated to this country.

Chrysanthemum

Pluck the petals from a giant-sized flower, rinse them quickly through running water, and sop off excess moisture. Top a vegetable salad, raw or steamed, with a liberal sprinkling of the petals. Spread them over soup, consommé, cooked potato and other vegetables. Include them in all cooked foods, even omelets. Dry and chop them for use in baked goods, pancakes, meat or fish patties, etc. Not only do they add extra color and a slightly pungent balsamic flavor, but they protect the system from lingering catarrhal and fatty deposits. This is especially true of three species: *Indicum, Segetum* (corn marigold), and *Balsamita* (costmary). The therapeutic power of the mum is a good reason to season one's *cooked* foods with it.

Chrysanthemum Soup

½ cup chopped scallions (or onion greens)
2 celery stalks chopped
2 tablespoons margarine or oil
5 cups chicken stock
2 carrots, sliced
1 tablespoon parsley or chives chopped
pinch of marjoram and/or basil
petals of a frech chrysanthemum flower

Sauté the scallions and half the celery in the margarine until soft, and gradually add the stock, stirring. Boil 1 or 2 minutes, add the vegetables and herb(s), and simmer for 20 minutes. Sprinkle the chrysanthemums over the soup and simmer 10 minutes longer. Stir and serve. There's but a brief whisper of the petal's balsamic taste.

Purple Coneflower
Brauneria pallida, B. angustifolia
Purple Daisy

The roots, fresh or dried, have been much used with barberry (which see) and other well-knowns, like sarsaparilla and dandelion, as an internal remedy for septicemia, boils, and other blood and skin problems. The old-time herbalist would say that since the large purple, conical head of the florets represented a boil, it was to be used for that and related blood disorders and to raise the body's resistance to infective conditions.

For home use, remove 2 or 3 complete root systems, wash clean, cut into small pieces, and dry them for a week. Reduce the dried, overground portion to a small size. Combine a tablespoon of this and a teaspoon of the roots, 1 tablespoon each of marigolds, verbena, and periwinkle, and ½ tablespoon of barberry, and vigorously boil the mixture in 2 pints of hot water for 15 to 20 minutes. Stir and strain. Drink ½ cupful every 3 to 4 hours. Sweeten to taste.

Coreopsis
C. lanceolata, C. coronata
Tickseed

In spring, strip the leaves from the plants and add them to a soup, stew, or chowder, or incorporate them, whole or chopped, in an omelet, hamburger, fish cake, or casserole. The Chinese call the plant *fang feng* and *t'un yun* and for centuries have used the young leaves as a nutritious vegetable. They employ the colorful flowers, seeds, and leaves as a post-illness tonic. The taste is sweetish; the aroma, faintly aromatic.

Use the *dried* flower heads to color wool and cotton an interesting yellow. You'll need 1½ cups of the flowers to dye 4 ounces of material. You'll find that the flowers of the appropriately named *Tinctoria* species yield a brighter shade. To this end, use alum with the first decoction, and cream of tartar to produce a bright yellow and chrome, a burnt orange.

Need a caffeineless pick-me-up? Mix together the dried flowers, seeds, and leaves, and infuse a heaping teaspoon in a cup of hot water. Use alone or with equal parts of bee balm, pelargonium, safflower, and/or carnation. Drink the tea tepid-warm 3 or 4 times a day.

Cosmos
C. sulphureus

This popular and showy member of the sunflower tribe is native to tropical America, Mexico, and the East Indies. In the latter region, the leaves and flowering tops of the young plants are a much-desired potherb. The somewhat peculiar, oily taste is modified by 2 or 3 minutes of steaming. The resultant yellow hue does well in rice, potato, and other starchy foods and may prevent the formation of what I call mucoid cement along the alimentary canal. You too may add these to soups, omelets, and casseroles.

The early Aztecs employed cosmos blossoms as a dye plant and paint. Prepare a yellow dye with the fresh yellow orange-rayed flowers, plus those of marigolds, crocus, and zinnia. Let 2 quarts of the flowers and one ounce of alum boil in a gallon of hot water for an hour. Strain, add the flowers, and boil 30 to 45 minutes. Remove and gently press out the liquid.

Flowering Crab Apple
Malus

September 1976. My son Irwin was showing me around the grounds of his new home and, of course, I could not resist suggesting his cultivating small herb patches here, a vegetable plot there, and to his ingeniously creative wife, Ruth, that she compote a few of the peach fruit "drops" ("and please save me the cuttings — the twigs and leaves — when you trim the tree"), and "Heavens to Betsy," I exclaimed, "look at that beautiful crab tree," and went into my usual pedantic spiel.

Did you know that the word *crab* is presumably related to the Scandinavian *scrab,* apple, in reference to the somewhat tart or mouth-puckering flavor of the raw fruit?

Did you know:

◆ That these fall-gathered apples, buried Indian style in sand or cold-stored for the winter, will lose much of their acidity?

Flowering Crab Apple

◆ That the enterprising homemaker, using conventional recipes, may convert the plentiful fruits into distinctively tasty jams, jellies, marmalades, and mixed preserves and pies? To flavor, you may add a judicious pinch of powdered ginger, cloves, cinnamon, or mint.

◆ That you can bake the crab apples and use them in pies, using honey to sweeten and powdered cinnamon, anise, or caraway to flavor?

◆ That you can pickle the fruits by covering them with sweetened cider vinegar for 2 weeks and use them with meats, fish, hors d'oeuvres, and sandwiches?

◆ That the fruits may be spiced? Warm a syrup of 1 pint of cider or herb vinegar and 3 cups of raw sugar, add 2 teaspoons each of cinnamon and cloves, and 4 or 5 allspice berries, cover, and let cool. Heat the syrup *slowly*, add the crab apples, and allow to cool overnight. Pack in clean jars, fill with syrup up to ½ inch to the top, and process in a hot water bath for 20 minutes.

◆ That the crab apples may be cooked in honey-sweetened ale, wine, or cider for 20 to 25 minutes and seasoned as above?

• That crab apple butter is a jam prepared by boiling the tart fruits in cider until reduced to a thick, smooth paste? Flavor with allspice and cinnamon and sweeten with brown sugar to taste, stirring well and often. Store in jars or crocks and cover tightly.

• That the rose-colored blossoms may be dipped in honey and eaten as a nibble like rose petals?

• That the acids of these fruits help to digest meats? For centuries British folk have cooked the tart apples with fatty meats such as pork and goose, the better to enhance digestion.

• That the fruits' tannin-rich juice helps to overcome summer complaints such as diarrhea? Two teaspoonfuls in ½ cup warm water taken every hour or two will suffice.

Cranesbill
Geranium
Geranium

Hearken, all ye that seek Nature's healing agents. Remember that this plant's exceedingly high yield of tannic and gallic acids makes it a fast-acting and safe astringent. The roots are nonirritating and lack any unpleasant properties. They are devoid of almost all taste, and so constitute a most desirable remedy for infants, for the elderly, and for the more fastidious.

In my early practice of herbal pharmacy, I prepared an herbal styptic that was 90 percent finely powdered cranesbill root. I'd discovered that the men of the 1800s had used the powder for all cuts, external ulcers, and hard-to-heal sores. And thus, I recommended it not only for shaving cuts but for bleeding gums and raw infectious wounds. For everyday use, therefore, gather the leaves before the plant seeds to obtain its greatest percentage of tannin, and boil a heaping teaspoon in 1½ cups of hot water for 10 minutes. Strain, and use tepid-to-cold.

You can prepare a gargle for the usual throat irritations and a soothing wash for mouth sores and thrush patches in children right in your own kitchen with your garden residents. Mix

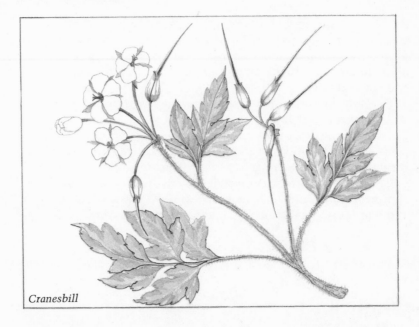

Cranesbill

equal parts of dried, cut roots and leaves of cranesbill, privet, yarrow, and loosestrife leaves. Simmer ½ to ⅔ tablespoon in a pint of hot water for 30 minutes or down to ⅔ the amount. Stir and strain. Gargle *warm* every hour. Refrigerate the balance and rewarm when needed. The slightly warmed liquid, diluted with an equal amount of tepid water, is an excellent injection for piles and leukorrhea.

To correct a temporary summer complaint (loose bowels or diarrhea), boil 1½ tablespoons of the above ingredients in a quart of hot water down to half (or for 20 minutes). Strain, and add a *small* pinch of cloves and cinnamon (or ginger) to each half cupful dose. Take warm every hour.

Crocus
Crocus sativus
Saffron

One evening in spring 25 years ago, I found Alan (son #4) in bed with a 103 fever and other symptoms of a bad cold. (Who ever heard of a good cold?) He had had only diluted fruit juices for the past 12 hours, yet was eagerly awaiting my herb remedy: a warm tea of mint and yarrow, with a touch of saffron. After only 3 cupfuls, one taken each hour, he was back to sleep. At 6:30 the following morning, his fever was down to 99. Unfortunately, he had also blossomed out with a gorgeous covering of measles! Nevertheless, from then on we seasoned various foods with the delightful herb saffron.

Saffron consists of the 3 deep red united or divided stigmas, which are borne on the ends of the slender elongated styles of crocus. For your purposes, gather both the stigmas and the yellow styles every October morning as the flowers open, and either pull them out or snip them with your fingernail. But be very careful not to lose the least tiny particle. Some 450 *dried*

Crocus

stigmas will level a teaspoonful. To make sure that the dew is almost completely dried off, collect them in the later afternoon of a warm day. Complete the drying process by spreading them out in a very thin layer on a sieve or window screen. Shake daily.

Saffron was in use for many centuries before Columbus set sail for spice-laden foreign lands. And here's how you too may enjoy this herb. To impart a golden yellow to wool, cotton, or silk, simmer ½ teaspoon of the *dried* stigmas in 4 cups of hot water, covered, down to half quantity or to the desired shade. Strain, add the material, and again gently boil and/or stir it in the resultant liquid. (Save and dry the vegetal remains for re-use.)

Want to add a slight buttercup lemon tint to your dull, lus-terless blonde hair? Simmer a heaping teaspoon of chamomile and marigold flowers in about a pint of hot water in a covered pot for 10 to 15 minutes. Add a 2-finger pinch of saffron and again simmer until a tablespoon of the strained decoction satis-fies your shade requirement. Meanwhile, shampoo your hair and dry off the excess moisture. Then, face down, vigorously brush your hair with the liquid, applying strength to the roots. Perhaps save and use again the remaining solution two days later? Refrigerate for safe keeping, and warm when needed.

After you've seasoned several of your foods with this all-purpose herb, you'll no longer wonder why its well-deserved reputation has lasted for tens of centuries. Homemaker, food connoisseur, and epicure alike agree: Saffron is an absolutely essential aromatic/coloring for starchy foods like homemade pasta, rice, steamed cabbage, baked or steamed potatoes, sauces, cream, butter, cheese, preserves, white wines, and such baked goods as bread, pastries, and cookies. (The botanically minded Will Shakespeare writes in *The Winter's Tale:* "I must have saf-fron, to colour the warden pies" Act IV, Scene 2).

For rice, stir well a little saffron in ½ cup of hot water to get a uniform strength and mix into the food, or add only enough of the powdered or finely ground herb directly into the pot. Do this 10 minutes before serving. For a potato, baked, steamed, or mashed, a few minutes before it's ready to serve, "dissolve" 2 or 3 of the powdered or ground stigmas in a teaspoon of near-melted butter and incorporate with the starchy portion. Rejacket a baked potato and note the oohs and ahs.

When a recipe calls for saffron in soups, stews, bouil-labaisse (fish chowder), fish or poultry, etc., use only *half* of the designated amount of that flavoring. Before using it, dissolve it in warm water (¼ teaspoon in ¼ cup).

A cup of rice needs ⅛ teaspoon of saffron. Soak the saffron in a tablespoon of warm water, add the rice, and cook for 20 minutes. Add the sautéed onion, pepper, tomato and mush-rooms, and cook 10 minutes more. Stir well.

In French, Italian, and Spanish cooking, the precious pow-dered or coarsely ground herb is often required to season/color chicken, but is a tablespoon to 3½ pounds of fowl really needed? Not necessarily so, especially in these inflationary times. Reduce to ½ teaspoon, add an equal amount of paprika and turmeric, and enough marigold and safflower to yield a ta-blespoonful. Make a curry powder of these five herbs, plus other powdered herbs and spices on your kitchen shelf.

Make use of the therapeutic benefits of this elegant aromat-ic. Stick with the claim that it is a very rich source of vitamin B^2, commonly known as riboflavin, and that its yield is about three times that of yeast or liver, ordinarily considered rich sources. This B complex member is required to prevent stomach and digestive problems, mental depression, nervous disorders, cracking at the corners of the mouth, etc. You too may use the herb not merely as a preventative of such prob-lems. The useful food aromatizer can become your gentle stomachic, nerve sedative, relaxant, and stimulant-tonic, and its diaphoretic (perspiration-increasing) property is especially needed in chicken pox and measles.

Try this simple remedy for any of the above-mentioned problems. Mix equal parts of yarrow, catnip (or mint), and lin-den, and a half part of sage, and a pinch of ground saffron. Steep a teaspoon in a cup of hot water and take every 2 or 3 hours. For feverish colds and oncoming skin eruptions, take every hour. The results, my four sons will testify, arrive within hours.

Next time you're bothered with annoying flatulent colic, stomach pains, or intestinal gripping, take advantage of saffron's spasm-relieving property. Drink a warm cupful of the above infusion every hour or two until relieved.

(It's interesting to note that Saffron-Walden, a town in Essex, England, was named after the plant. For over 200 years large quantities have been cultivated there.)

D

Daffodil
Narcissus pseudo-narcissus

Want to make your own facial using a recipe that's been around
for nearly 2,000 years? Finely powder equal parts of *dried* daffodil bulbs, barley, and lentils, and make a paste by adding
water and honey. This is the time-tested recipe of the Roman
poet Ovid who claimed that "every woman who covers her face
with this will make it more brilliant than a mirror." And in
Elizabethan times, the mixture, minus the lentils, was also
used to heal assorted skin eruptions and discoloring; the roots,
boiled in oil, were (and are) applied to raw, rough-skinned
heels; and to heal an infectious spot on the ear, the bulb's expressed juice was mixed with honey and an equal amount of
wine (tincture) of myrrh and applied to the ear.

If, while gardening, you suddenly acquire a bruise or sprain,
a thorn or felon, pound or crush a small bulb and apply it and
the juice to the affected area. Better to bandage the mass. A
second application may be needed to remove the attending inflammation and pain.

Having trouble with moles who just love your tulips and
gladiolus? Daffy bulbs are known deterrents when planted
around the bed you want to protect.

Dahlia
D. variabilis

Bravo! You're quick on facts concerning this plant. You know
that it was named after Dr. A. Dahl, a famous Swedish bota-
nist, and that its original home was Mexico's high moun-
tainsides. You know that dahlias require humus-rich soil, warm
weather, and full sun, and that they bloom in late summer.

But how to make use of this garden plant commonly found in our more southern states? By eating it, of course. You'll find that the tubers, steamed or baked, are easily digested and a source of needed nutrients, especially inulin (not insulin). Everyone can eat them, whether they are sick or well, without fear of harmful aftereffects. And try adding the fresh, colorful petals of the dahlia to all your salads, soups, and stews when serving.

From the inulin, which is also found in dandelion, chicory, and elecampane, there is produced pure levulose, which is known as fructose and diabetic sugar. Studies have indicated that it may have special utility in the treatment of diabetic patients when parenteral carbohydrates are required as a result of surgical or medical complication. Another advantage: Given intravenously, it allows a higher concentration of sugar to be administered, thus providing more calories more rapidly without producing hyperglycemia, glyccsuria, and diuresis.

Therefore, all, especially sugar-avoiding diabetics, may enjoy the sweet and pleasant taste of the cooked tubers. It's interesting that the candy industry uses powdered extract as a retarding agent in the manufacture of crystallizable sugar products and that the brewing and mineral water industries use it too.

At one of the Scottish universities during World War I, an extract of dahlia bulbs proved to be a most valuable and efficacious medicinal, i.e., a health-restoring tonic for recuperating British soldiers. And in European sanitariums, the levulose extractive is given to tubercular children and others afflicted with wasting diseases.

English Daisy
Bellis perennis
Bruisewort

Call this plant by its synonym, bruisewort, and you'll realize that it too is a most dependable remedy for quickly healing most external pains and aches, wounds and skin irritations. You'll find that a strong tea or mild decoction (1 to 2 teaspoons to a cup of hot water) helps gout, scorbutic and feverish complaints, if taken 3 or 4 times a day for at least 10 days in combination with a diet that is near-vegetarian. Take it too for liver complaints.

Include a few of the somewhat sharp-tasting young leaves in your soup or casserole, meat or fish patties; or cook them with beans, peas, broccoli, squash, and cauliflower.

Oxeye Daisy
Chrysanthemum leucanthemum

Here's a kitchen medicine for a painful bruise, strain, or swelling: Pour hot water in a pot, enough to cover 2 cupfuls of the leaves and flowers of this plant. Cover 5 minutes and apply as a

Oxeye Daisy

hot fomentation every hour or two as required. Massage the affected area with the liquid. Cooled, it may be patted onto sores and skin eruptions. For the latter purpose, ½ cup of the flowers, dried and ground, may be slowly simmered in a cupful of vegetable oil, melted lanolin, or unsalted lard for 10 minutes. Stir occasionally. Strain and label the ingredients. Added ingredients to both remedies: English ivy, hound's-tongue, and water lily.

Let the finely powdered flowers serve you as an insect repellent. Incorporate 2 parts each of oxeyes and (powdered) arborvitae and one part each of borax and sulfur and spread the powder thinly over the threshold and wherever the unwelcome pests may appear. The daisy's active substance scabrin has been duly recognized by the U.S. Department of Agriculture as a deadly weapon against army worms, flies, and other insects.

Day Lily
Hemerocallis fulva, H. flava
Lemon Day Lily

August 1953: James Lee, then a student at Worcester Academy, told me that folks in his native China used a great deal of red clover and other wild plants as well as garden-cultivated ones like day lilies. Oriental cuisine, he said, required that the unopened buds first be dipped in hot water, chopped, and taken with salads.

June 1972: Margaret ——, a mother of an adventuresome four-year-old phoned me to ask what antidote to use for day lily. Her son had eaten several flower buds and she was quite worried. ("Frantic!") Should she rush him to the hospital? She did not expect my answer: "You and other members of your family should eat the buds and the opened flowers too — in your salads, with soup and other cooked foods. I wish all garden plants were as edible as the day lily."

Here are a few ways of getting better acquainted with this hardy perennial. Near June's end, gather the green-tongued flower buds or, later, the fully expanded flowers and soak them quickly in water before using. Chop and mix them into a soup, stew, or fish chowder, already cooking, to yield a gumbolike effect; or incorporate them, minced, in an omelet, casserole, or a fricassee of chicken or veal; or sauté with noodles, vegetables, rice, or buckwheat. The closed, shriveled blooms may be similarly used. The water-heated flowers and buds offer a bland but interesting taste.

Flower Pancakes

Mince or use 2 whole flower clusters. Dip them into a batter consisting of ½ cup of whole-grain flour, 1 egg, ¼ cup of milk, and a teaspoon of honey. Brown both sides in oil or butter, like fritters, and garnish with butter or maple syrup. (Thanks to Blanche Derby for this recipe.)

If you're going to use the *fresh* expanded flowers, do remember that they're open for culinary business only for one day. (Now you know that *Hemerocallis* means that the flowers are beautiful for a day since the blossoms close at night.) Why not refrigerate a few if you're going to use them within a few

Day Lily

days? Otherwise, the excess may be saved for future use by drying. Gather the buds and flowers and dry them on unmarked (printless) cardboard in a room or attic where warm air circulates. Complete drying takes 5 or 7 days, depending on weather conditions. Store the buds and flowers in separate sealed containers.

And those clusters of fleshy tubers! In spring, even before the stalks arise, gather the firm ones by separating them from the main root. Wash-clean and steam, bake or cook them, whole or sliced, with other vegetables. You'll enjoy their near sweet corn taste.

To add extra flavor or remove their "crude" flavor, soak them in herb-flavored (i.e., marigold, lavender, carnation) wine or vinegar for 2 or 3 days.

Pickled Buds

a quart of buds
½ pint garlic or afore-
    mentioned vinegar

2 teaspoons dillseed
2 teaspoons basil

Cover the buds with water and boil 1 to 2 minutes. Drain and pack into sterilized glass jars. Dilute the vinegar with half as much water, add the dillseed and basil, and boil for 5 to 6 minutes. Pour the mixture over the buds, seal, and let stand 2 weeks.

Dictamnus
D. albus
Gas Plant, Dittany, Fraxinella, Burning Bush

Rub any part of this fragrant plant and you'll discover its lemony odor and guess its healing properties. To use, gather the upper half of the plant and dry by suspension.

You don't have to be ill or even temporarily out of sorts to enjoy drinking a tepid infusion of the dried leaves and flowers (a heaping teaspoon to a cup of hot water). But should a jiffy remedy be necessary for colic pain, a nervous stomach complaint, or a fast-acting sweating agent for a fever or cold, then use the same warm tisane as often as needed.

Flowering Dogwood
Cornus florida
Boxwood

If Hippocrates considered the parts of this shrub as near cure-alls, so may you and I. A chronic problem of dysentery may be treated by boiling a level teaspoon of the dried berries in a pint of hot water down to half the liquid content, straining, and drinking it warm every hour or two. For a fever or lasting cold or weakened condition of the stomach, mix equal parts of the leaves and twigs, boneset or sage, mint, and safflower and drink a hot infusion (a teaspoon to a cup) every hour. And with the addition of equal parts of the shrub's flowers and bark and

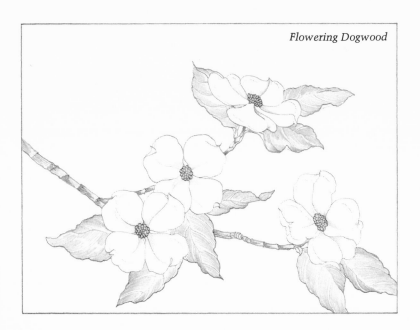

Flowering Dogwood

chamomile to the resulting tea, you've prepared a good healing poultice for ulcers and acute skin infections.

Make your own healing ointment by simmering all *dried* and finely ground parts (leaves and larger twigs), yellow dock, and Solomon's seal roots in hot lard and lanolin (a handful to a cupful of base) for about a half hour. Stir occasionally and strain while still warm. Use this good antiseptic/astringent ointment in all minor skin complaints.

Did you know that you can use the bark-stripped twigs as a dentifrice just by chewing them and rubbing them against your teeth? That the juice of the twigs helps to harden the gums?

E

Eupatorium
E. purpureum
Joe-Pye Weed, Purple Boneset

Will wonders never cease? This plant's name is said to be derived from Mithridates VI Eupator, c. 63 B.C., King of Pontus, who successfully used the plant's juice as an antidote against a variety of poisons. Thus a mithridate — a sovereign and almost panacean remedy — will protect you against nearly all organic disorders.

Dry the entire overground portion and reduce to tea size. No matter what ails you — a feverish cold or sore throat, gout or rheumatism, or especially chronic kidney or bladder (cystic) affections — you'll find positive and lasting relief by drinking a warm tea of the leaves and flowers every 3 to 4 hours (a teaspoon to a cup of hot water). Food must be restricted to fresh fruits and leafy vegetables at the same time.

If you're a butterfly collector, have your net and/ camera close by your patch of purple eupatoriums in September, when the flowers' fragrance and magenta odor attract a large variety of butterflies.

F

Feverfew
Chrysanthemum parthenium

When you're gathering the fully matured feverfew plants, you can't escape its disagreeable scent. Nibble the leaves or petals and you'll never forget its decided camphoraceous taste. There is an old-time trick of keeping bees and flying insects away by attaching a few flower heads to your hands and hat.

Know that its name is a corruption of febrifuge, to expel fever, and you may want to use the strongly scented over-ground portion to treat feverish colds. For this purpose and for any gastrointestinal upset, catarrhal problem, or simple nervous disorder, sip a warm infusion of the flowers and leaves (a teaspoon to a cup of hot water) 3 or 4 times a day. Or steep a teaspoon of feverfew with equal parts of verbena, lilac, linden, and marigold.

Flax
Linum usitatissimum

You don't have to be an Egyptologist to know that flax was commonly used in Israel and Egypt and in Central Asia thousands of years ago. While I do not expect you to prepare linen from the stemfibers or to express oil from the seeds, you *can* include the seeds in a gumbo soup, stew, or casserole. You can roast them as a nibble. You can use them powdered in bread baking. And you can mix them into your bird's food. Their protein content is enormously high (20 percent), as is their sugar content (18 percent). And you'll get used to the okralike, oily taste.

You may want to know that soaking the flat, brown seeds in hot water produces a gummy extract. Should you need a soothing demulcent for a raspy cough and bronchial problems, as well as gastrointestinal, urinary, and catarrhal conditions, boil 2 or 3 teaspoons in a pint of hot water down to half the liquid and take it, tepid-cool, in wineglassful doses (3 to 4 ounces). Temporarily constipated? Incorporate 3 to 4 teaspoons of the ground seeds with enough of the liquid to fill the cup, and drink one such cupful morning and evening, followed by another cup or two of tepid water or herb tea (bee balm, mint, carnation). And to normalize a temporary malfunction of the kidneys, steep a teaspoon each of flaxseeds, the leaves of mint, eupatorium, and hollyhock in a cup of hot water. Drink the tea tepid-cool 4 times a day.

Perhaps your need is a poultice for a boil or particular skin infection, bruise, or sprain? Cook a cupful of the whole (or pulverized) seeds in 2 cups of hot water until they're soft and gooey, enclose them immediately in cloth or linen, and apply as hot as possible.

Forget-Me-Not
Myosotis

Were my grandfather to see this plant in your garden, he'd say without hesitation: "Know why it's so good for insect bites, stings, and other skin itchings? Observe that the curled flower head resembles the tail of a scorpion or small snake. Soak the dried, mature plant in rubbing alcohol for 10 to 15 days and strain; or cook a tablespoonful in 2 ounces of melted unsalted lard for 20 minutes and strain into an ointment jar." Label both items with the ingredients, date of preparation, and their intended use.

For that annoying dry cough or deep-seated bronchial affection, prepare a syrup by simmering a handful of the dried leaves and flowers in a pint of hot water down to half quantity. Let cool and strain. Then add enough (raw brown) sugar or honey to thicken. Sip a tablespoon slowly every hour and that hardened mucus will be loosened and expelled from the respiratory and bronchial organs. In 1964 an English herbalist told me that this very syrup is still prepared "back home" and prescribed for pulmonary (lung) affections and emphysema.

Forsythia
F. suspena

Welcome this early flowering bush, this true harbinger of spring, to the club of valuable and useful ornamental plants in your garden. Perhaps you can't prepare as many remedies using the forsythia as with other plants listed elsewhere. It is, however, an ever-present and rich source of a healing bioflavonoid called rutin that is closely related to vitamin P. Other natural sources are buckwheat, spinach, rice, red currant, tomato, pansy, elder, hydrangea, and rue (whence rutin).

I am not saying it's a cure-all, rather that if we—your family and mine—ate some of these foods and drank daily health teas, we'd find that their rutin content would help to prevent internal bleeding due to capillary fragility and so markedly reduce a possible incidence of recurrent hemorrhages. A glance into medical books will inform you that rutin is an effective preventative for persistent nosebleeds and bleeding gums and is often prescribed by physicians in the treatment of a disease like diabetes, where there's a possibility of the rupturing or weakening of the capillary system.

Rutin therapy has been medically recommended for patients with hypersensitive cardiovascular problems and found especially efficacious in lung hemorrhages, where previously a series of blood transfusions was required. The same has held true in cases of bleeding gastric ulcers.

More importantly, medical researchers have recently observed the close relationship between vitamin P (rutin) and the adrenal cortex. The vitamin is invaluable in helping to overcome the ill-effects of old age and premature deaths of older folks, which may be due to capillaries weakening or completely breaking down.

Forsythia

Members of my herb-study classes have expressed praise for health teas made with forsythia. In time you will too. Gather the flowers in bloom and the summer leaves and dry and grind or cut them. Stir a heaping teaspoon in a cup of hot water, cover 15 minutes, and strain. Drink one such cupful 2 or 3 times a day—upon rising and between meals. Sweeten with honey, if so desired.

Fringe Tree
Chionanthus virginicus

This small ornamental shade tree bears delightfully fragrant white flowers in early summer. "What a pleasure to rest 'neath its spread," Susan L. would often say. "How soothing and comforting to a troubled mind and body, especially for one in low spirits. If you're fortunate enough to find such a haven, pause

there a few minutes, especially on the second dry day following
a rain. Sit or lie down, close the eyes, re-e-e-lax, inhale deeply.
A few minutes of such meditation is a superb restorative to all
your senses, an equitable pick-me-up that invigorates the entire
body."

For another kind of pick-me-up, mix a heaping tablespoon
of the dried, cut twigs (summer collection) of the tree, of dogwood and hibiscus, the cuttings of barberry, the leaves of
eupatorium, and a teaspoon of gentian (which see). Suitable re-
placements: the leaves and flowers of bachelor's button and
dictamnus. Simmer *slowly* a tablespoon in a quart of hot water
for 15 to 20 minutes, stir in 2 tablespoons of mixed aromatics
(mint, alyssum, and carnation), and cover until tepid. Strain and
sweeten with the least honey or sugar. Take ½ cup (with or
without added water) 3 or 4 times a day, an hour before breakfast, between meals, and an hour before retiring.

You'll find that this preparation or an aromaticized tea of
the twigs alone is especially good in acute dyspepsia complaints
of the liver and gallbladder because it liquefies the bile and
hopefully prevents gallstones (*if* the diet is respectfully guarded
from most man-made foods). Drink the tea as well to remedy
all stomach disorders, except constipation.

Collect the twigs (now dry) in fall, and you will have a
quickly healing and astringent application for your cuts,
wounds, and minor skin diseases. Prepare a concentrated lotion
by strongly decocting a cupful of the ground twigs — use the
bark of the lower trunk too — in a quart of hot water until a little less than half the liquid remains, cover until cool, strain,
and sop onto the affected area. Apply every ½ to 1 hour as is
convenient.

You're very fortunate if you have access to the purplish
fruits of this low shrubby and ornamental tree. Gather the ripe
ones in midsummer and preserve them in cider or herb vinegar.
Serve them as an extra pickle, condiment, or with a relish to
accompany meat or fish.

G

Gentian
Gentiana lutea
Yellow Gentian,
G. purpurea
Blue Gentian

During the years 1930 to 1950, my older Swedish customers would present me with a kind of glogg, a highly spiced and highly alcoholic wine. These share-the-wealth folks, knowing of my interest in herbal matters, revealed that the major ingredients of the liquor were their plentiful whole gentian and hollyhock plants—flower, leaves, stem, and root, and hawthorn fruits.

Swiss old-timers have also suggested using gentian, wormwood, yarrow, chamomile, and citrus rinds as the base for a vermouth-type wine. Usual standard formulas will do, and the finished product, you'll find, serves in several ways. The homemade brew, of course, may be taken as an aperitif or in mixed drinks. This herbal tincture is considered a most invigorating tonic for young and old. Feel out of sorts, cold, or just below par? Mix one or two teaspoons in ½ cup of tepid water; slowly sip the amount 3 or 4 times a day. You'll also find it's a quick-acting remedy for most stomach disorders: It tones the gastrointestinal system, restores normal appetite, and assists digestion.

Include that weedy nondescript dandelion in the wine's preparation, and your product somewhat represents two once popular drugstore medicines, Elixir Glycerinated Gentian and Gray's Glycerine Tonic. And did you know that gentian root (or its extract) gave the sharp "bite" to Moxie, a carbonated beverage that was very popular a generation or two ago?

Make a good marinade of your recently finished liqueur. Mix 2 cups of the wine, ½ cup of malt or cider vinegar, ¼ cup of vegetable oil, 1 clove garlic (chopped), a large onion (chopped), 1 teaspoon of mixed herbs (basil, savory, rosemary, oregano, marjoram), and 1 tablespoon of dried rinds of lemon,

orange, and/or tangerine. Pour all ingredients into a large pot or
roasting pan, warm at the lowest heat, add the food to be mari-
nated, cover, and cook for 10 minutes. Warm-baste the food
every 10 to 15 minutes.

Why use the marinade? Because it predigests coarse
fatty meats and fish (and venison) and removes the harmful
cholesterol-building fat. And it produces food that is delightfully seasoned and temptingly tasty.

Strain and save the marinade for a future hot-basting liquid
and preseasoned gravy or sauce. When preparing other meats,
rewarm the leftover marinade for a baste by adding a sprinkling
of the same savory herbs and ½ teaspoon gentian root contained in cloth.

Feeling out of sorts, downright blue? If you're an adventuresome soul, you may take my gentian-containing standby of
many years. (I borrowed the idea from Grandfather's wine formula and from an eighteenth-century compound called Baum
de Vie — "Elixir of Life" — prepared by Le Lievre, a French
apothecary.) My product consists of equal parts of the C
spices — powdered clove, cinnamon, and cardamom — and a

Gentian

quarter of their total of powdered gentian. Stir ¼ teaspoon in ½ cup of hot water or in a cup of mint, basil, or other herb tea, and drink 4 times a day, after 2 hours of not eating. It's a good restorative tonic and a reliable corrective in most digestive and intestinal complaints, in flatulence, and in diarrhea. Many of my older customers considered it a reliable antirheumatic remedy.

Do save and dry the overground portion of your autumn gentians for your own bitters or spirits-reviving tonic. Reduce the parts (leaves and stems) to a coarse powder, combine with an equal portion of ground yarrow, birch twigs, rose geranium (pelargonium), bee balm, and other kitchen aromatics, and let the mixture soak in wine for 10 days. Shake the bottle once a day. Strain and take an ounce in tepid water every 3 or 4 hours.

(If you're going to use gentian root, collect it in late summer and allow it to stay on the ground for some time and ferment.)

Geum
Geum

Enough! These hardy perennials have served you well as border guards of your flower garden, but there are other ways they may fetch you greater profit.

Dig up a root of the *Urbanum* species and smell the clovelike aroma. That's the clue that led folks in Grandfather's day to chew the root for offensive breath. And when you're readying your home brew — beer or wine — add extra flavor with a handful of the spicy, dried roots.

Use the dried root, cut, as a tonic-aromatic for stomach disorders and as a fever-chaser in colds. An infusion of ½ to 1 teaspoon to a cup of hot water is taken warm every 2 hours. And cold, it's a good astringent in summer complaints and sore throat.

A Good Spring Tonic
Mix equal parts of the dried roots, verbena, spirea, bittersweet,
and hydrangea. Simmer a tablespoon of the mixture in a pint of
hot water down to half. Strain and take a tablespoon, tepid to
cool, in ½ cup of water 3 or 4 times a day.

Another Version
Macerate an ounce of the ground root in a pint of unsweetened
white wine for 2 weeks. Add 2 teaspoons each of carnation,
barberry, verbena, and spirea, and ½ teaspoon of cinnamon
(sticks), mint, and New Jersey tea. Cap the bottle and let stand
another week. Shake the bottle every other day. Stir and strain.
Take 2 to 3 teaspoons in ⅓ cup of water 3 to 4 times a day.

And if perchance you should note the recommendation of
Culpeper and other Elizabethan herbalists that a wine or infusion of the roots is a "good preservative against the plague or
any other poison," smile if you will, but be aware that aqueous
extracts have been found effective against the pathogenic staph
organism that is implicated in boils, abscesses, and suppurating
wounds.

Gladiolus
G. communis, G. illyricus

Rank superstition? Old wives' tale? Need we believe, as the ancients did, that this plant's physical appearance foretold its use
in therapeutics as well as its stimulating and aphrodisiac properties? That since the blades symbolized the male organ, the
lower leaf section and upper half of the root would serve as a
"provoker of love"? Today its sword-shaped leaves should remind you of the plant's tonic and "gladiatorial" qualities (from
the Latin *gladius*, "a small sword").
To prepare a mildly stimulating body toner, gather the
erect blades at flowering time, rinse them thoroughly, remove
the excess moisture, and cut or chop them into ½-inch lengths.

Simmer 2 heaping teaspoons in a pint of hot water for 15 to 20 minutes, strain, and let cool. Drink ½ cupful upon rising, between meals, and 2 hours before retiring. Add, if necessary, dictamnus, mint, or other aromatics to season, and honey, to sweeten.

Prepare a healing lotion for cuts, scratches, and skin affections by concentrating (boiling down) a decoction of the dried roots (a large handful to 1½ pints of boiling water) in a covered pot. When the water content is half, let cool and strain. Sop the liquid onto the affected area as often as needed.

H

Hawkweed
Hieracium
Orange, Mouse-Ear

Associate this flowering plant with birds, and particularly with falconry, and you'll never forget its purported benefits to the eyes and liver. First in Eastern lands (Iran and China) and then in medieval England, hawks (falcons) were observed to consume parts of the hawkweed; and their sight so sharpened that their owners, the falconers, became aware of the plant's healing powers.

The plant was found to be of equal benefit in improving and strengthening human eyesight. Perhaps the falconers too tried eating the flowers and leaves or drinking hawkweed in their tea. Yellow- and orange-flowered plants like the hawkweed have long been used by herbalists in preventing disorders of the liver; and most likely hawkweed ensures the use of vitamin A in the liver, a much-needed factor to good eyesight.

Your doctor may tell you that eye weakness is often associated with or is caused by liver derangements. So, to protect your liver and improve and protect your eyesight, do try to eat lots of orange-colored foods and deep green vegetables that contain high quantities of vitamin A; and do drink herb teas of hawkweed, dandelion, safflower, and other similar yellow-flowered plants. Dilute ¼ cupful of the tepid infusion (a teaspoon of the finely ground herb in a cup of hot water) with an equal portion of tepid water and drink it slowly 3 or 4 times a day, eating nothing else but a few sweet grapes or other sweet fruit.

Use the same tea for bilious attacks, gallbladder malfunctions, jaundice, "spots-before-the-eyes," and liver problems. Dilute an ounce of the originally infused tisane with 2 parts of warm water and take morning, afternoon, and early evening. And eat very little except a bit of fruit. The tea is good too for the nonailing.

Hawthorn
Crataegus oxyacantha

You're a home conservationist and you don't like the taste of the haws (fruits) of this shrub — think they're rather insipid — and you've never used them in any way whatsoever. Willing to try?

True, they're not overly palatable until the fully mature fruits are bright red and gathered in midfall. Even then they are admittedly not first choices. However, although they contain large stones, the pulp is fairly juicy and can be used in conventional recipes for jellies and marmalades, with the addition of orange, lemon, and/or grapefruit peels. Season with savory, thyme, marjoram, etc. You'll find that hawthorn preserves require comparatively little sugar.

You can make a zesty pickle of fruits, alone or combined with crab and sour apples, assorted berries, cranberries, etc. Bring to a quick boil 1 pint cider vinegar, 2 cups raw brown sugar, and ½ cup water. Add 2 pints of the fruits, slowly bring to a boil, lower the heat, and simmer 5 to 7 minutes. Remove from the heat, and add a cloth bag containing 2 tablespoons of mixed pickling spices to the mixture. Place fruit in jars with the syrup poured over. Immerse overnight and then seal.

Use standard recipes to prepare a fruit relish and chutney with hawthorn fruits as the main (and free) ingredient.

Don't neglect the white flowers. Gather them in May and June and use them dried in this "sedative" remedy for temporary nervousness and insomnia: Mix equal parts of lady's slipper roots, catnip or mint, chamomile, and the flowers, and steep a teaspoon of the mixture in a cup of hot water.

Have need to normalize blood pressure? Go to bed and fast (drinking water only) for 2 days, then take an herb tea of elder and hawthorn flowers, chamomile, and lavender (a heaping teaspoon to a cup of hot water) 3 to 4 times a day. Eat as sparingly as possible for the next day or two until well recovered. *Or:*

a. The above four ingredients, powdered, may be taken 4 times a day in #00 gelatin capsules, obtainable at your pharmacy.

b. Soak a heaping teaspoon of the crushed hawthorn fruits

in ⅔ cup of cold water overnight. Bring to a boil, simmer 10 minutes, and strain. Drink 2 or 3 ounces every 3 hours. If necessary, sweeten with honey.

It's interesting to note that an extract of the hawthorn fruits sometimes appears today in prescription drugs to maintain a lower level of blood pressure brought about by other medicines, thereby reducing the patient's constant chest pains and recurrent headaches.

For that reason, it is also used in treating coronary and diuretic problems. A hawthorn-containing drug item, now manufactured by a Long Island, New York, pharmaceutical concern, is often prescribed by physicians for the symptomatic treatment of essential hypertension "to help reduce blood pressure quickly, to relieve accompanying headache, dizziness, and to effectively prolong the reduction for treatment of basic causes." Its active constituents, *crataegin* and *amygdalin*, have been used in medicine as a heart tonic and to sharply reduce the grievous symptoms arising from urinary and cardiac conditions.

Heliotrope
Helotropium arborescens

If this interesting "sunflower" now occupies a sunny window location in your home or a spot along the warm, moist border of a garden, you can luckily gather the clusters of the small, richly colored, reddish purple flowers. And how to use them, you ask?

Use them as cut flowers for the table. (They'll last longer if you add 2 ounces of prepared tea and a teaspoon of glycerin to 8 ounces of water.) And at long last, you've noticed the plant's sweet vanillalike scent? The fragrance arises from rather modest flowers and may denote a moral of sorts. Folks with a most pleasing, "fragrant" personality are often the most humble, modest, and unpretentious.

Do use the colorful and fragrant flowers in sachets. Dry

and strip them from the stems before incorporating with other ingredients. (See Violet.)

For distressing discomforts of menstrual periods, infuse 2 heaping teaspoons of the dried, ground flowers in a cup of hot water for 10 minutes. Strain and stir in ⅛ teaspoon each of powdered ginger and clove. For minor uterine discomforts, simmer a level teaspoon each of the flowers and mint and/or catnip in 2 cups of hot water in a covered pot for 10 minutes. Strain and drink warm every 2 hours as needed. (Tincture of heliotrope has been prescribed for the latter problem by homeopathic physicians.)

Canadian Hemlock
Tsuga canadensis

This tree often attains a height of 70 to 80 feet and thus acts as a windbreaker and source of shade. That's good, but do consider it an excellent source of Canada pitch with which to prepare a skin salve. Here's how:

Gather the resinous sap that exudes spontaneously and dries and hardens on the bark. Boil the bark and pitch in hot water for a few minutes (that was Grandpa's method), skim off the pitch, simmer a tablespoon in 2 cups of either unsalted lard, lanolin, or suet (equal amounts) until the gum is all dissolved. Strain into suitable containers and label. The ointment resulting quickly heals and relieves the pain of pus-filled cuts and sores.

Hepatica
H. triloba, H. acutiloba, H. nobilis, Anemone hepatica
Liverleaf

The common synonym for this plant, liverleaf, refers to the shape of its nearly evergreen leaves, which are angular and three-lobed and in the eyes of the beholder either kidney-, heart-, or liver-shaped. The ancients named it well, and correctly, after the Greek *hepar*, the liver, or *hepatikos*, affecting the liver.

You'll find that the leaves and flowers make a demulcent drink in liver or kidney disorders, indigestion, and bronchial affections. Slowly simmer an ounce in a pint of hot water for 20 minutes, strain, and drink ½ tepid cupful every 2 hours or as needed.

A Catarrh Remover or Cough Remedy
Blend equal parts of the flowers and leaves of violets, hibiscus, and veronica, and steep a tablespoon in a cup of hot water. Drink the tepid strained liquid every hour or two.

Hepatica

Relieve the pain of your recently acquired bruise or sore spot: Apply a hot poultice of the entire *dried* plant, with or without equal parts of wormwood, tansy, or sage.

Prepare an effective emollient facial by applying a warm wet pack of the cut leaves and flax leaves and seeds to the skin. Helps to soothe out wrinkles, and reduce the irritation in acne, pimples, and other skin problems.

Hibiscus
H. syriacus
Shrubby Althaea, Rose of Sharon

Uh, uh. Please don't throw those trimmings away! If you must prune and cut the limbs back because of this late-blooming shrub's spreading, at least do save the leaves. Reduce the twigs to a much smaller size and use as part of a mulch or compost pile.

Its generic name is derived from the Latin word meaning "large" or "marshmallow" (whence the synonym shrubby althaea), and should remind us of the profitable uses of such relatives as hollyhock, okra, the mallows, cotton, and almost two hundred others. The early leaves may be steamed with other vegetables, used in soups or stews, diced and included in casseroles and omelets. They may also be cooked with Irish moss when preparing thin jells or a pudding. Hibiscus leaves give these preparations a gumbolike consistency. It's their mucilaginous content that thickens the cooked items and helps to allay the pain of internal irritations.

The later leaves are a suitable replacement for mallows or hollyhock, both as a food and as a healing agent. When my sister Annie and I were young, folks depended on the hibiscus for a constant supply of leaves for their cough remedy. How to make it? Grind the dried leaves until they are greatly reduced in size and simmer a heaping tablespoon in about a pint of hot water for 15 to 20 minutes. When cool, stir and strain, expressing as much of the liquid as possible. Add enough sugar or honey to sweeten and take in tablespoon doses every hour or two, or as needed.

Holly
Ilex aquifolium

This handsome evergreen has played an exciting part in folklore and medicine, science and history.

Holly has had a long history of medicinal values, first attaining prominence two thousand years ago. Its properties are well described by the Roman writer Pliny and the Greek healer-physician Theophrastus. The former claimed that a holly tree planted near one's home protects the house from lightning and evil witches. Most revered in biblical days for its supernatural influences and medicinal virtue, the holly was called Holy Tree, a name also identified with holm, *Quercus Ilex*, an

Holly

evergreen oak with hollylike leaves. Even an old Christmas carol alludes to holly:

> *"Christmastide*
> *Comes in like a bride,*
> *With Holly and Ivy clad."**

The custom of decorating homes with holly at Christmas is said to have originated with the Roman custom of sending boughs to friends during the festival of the Saturnalia (which began on December 17). Herbal folklore, according to M. Grieves, credits the Druids with the custom of decorating their houses "with evergreens during winter as an abode for the sylvan spirits."

Cut down to size, the holly is often utilized as a hedge. Gather holly leaves and young buds about noon on a sunny day in May or June and, when they've completely dried, reduce them to a smaller size (with a coffee or spice grinder). Now put them to good use. If your ailment resembles "wandering rheumatism" or temporary kidney dysfunction, take the suggestion of eighteenth-century herbal physicians: Drink a strong honey-sweetened holly tea (2 teaspoons of dried leaves and young buds to a cup of hot water) 3 or 4 times a day, and try to abstain from solid foods.

Use the same parts of this plant for their emollient and expectorant properties in treating deep-seated coughs and catarrhal affections. Steep a heaping teaspoon of a mixture of equal parts of holly, hollyhock, and thyme in a cup of hot water and drink a tepid cupful morning, afternoon, and mid-evening.

Hollyhock
Althaea rosea

Hollyhock is one of my favorite garden tenants. For the little space it takes up along the foundation of my home and the

*Mrs. M. Grieve, *A Modern Herbal*, p. 405

stony area next to the garage, it pays excellent rent and yields good economic returns, as do my other tenants — barberry, mallow, rose, wormwood, and raspberries. I've made excellent use of the plant, which many of our ancestors called "outlandish rose," in a variety of recipes and healing remedies, as a soil (organic) enrichener, and mulch material, etc. . . . And so should you.

A Bit of History

This handsome guardian of the portals (at my house anyway) has been cultivated, and in constant use, by the people of China and India, sometime before history began. At a time when horticulture was little understood, the herb was grown (as it is today too) in vegetable gardens *as a food.* The Crusaders introduced it as a potherb to England and Europe and christened it holy hoc(k), from the Holy Land. Along came the early colonists to American shores and so did hollyhock seeds.

The young leaves are edible and an excellent money-saver. Gather them freshly in spring and before adding them to a vegetable salad, let them soak in an herb vinegar for 3 or 4 minutes. Preseasoned and cooked with other foods, they're a culinary delight. Steam or cook them with vegetables and add them to all casseroles and soups, especially Creole gumbo. When you're preparing lentils or baked beans, add a tablespoon of diced leaves to each cupful of either food. And when fritters of fish, meat, poultry, or vegetables are in order, make the finely minced leaves part of the batter. Substitute them for grape leaves.

Hollyhock is related to hibiscus and can be similarly used. Even consumed as food, it has healing effects, emphasizing the origin of its generic name, Althaea, which is from the Greek word *altho,* "to cure."

Whenever you're making a herbal remedy for any internal organ of the body — the bronchia, kidneys, or gastrointestinal area — be sure to include the dried, cut, or ground parts of the hollyhock. The leaves, flowers, and roots are well known for their demulcent/emollient and healing properties, helpful in removing harmful catarrhal and gravel deposits.

A Health Drink
Steep a teaspoon of the dried flowers and leaves of rose, mint, and hollyhock in a cup of hot water for 10 minutes. Try a cupful too in place of your usual Pekoe tea.

A Cough Remedy
Simmer 2 teaspoons of the leaves and a teaspoon each of thyme and mint in a pint of hot water down to half the amount. When cool, strain and make a syrup with enough raw sugar or honey. Sip a tablespoon slowly every 2 hours.

A Kidney or Stomach Remedy
Prepare as above, using a teaspoon of a mixture of yarrow, pelargonium, and marigold. Omit the sugar. The dose is the same.

Want a substitute for litmus paper — to test for acids and alkalies? Cut uniform slips of white filter paper and dip them into a strong infusion of the dried flowers to render a permanent purplish blue color. If you add an acid to the liquid, the color turns red; add an alkali, bluish green.

Make hollyhocks a garden favorite of yours. Enjoy their indispensably elegant and decorative effects, their variety of colors, and especially their easy cultivation. Besides the above-mentioned benefits, they are useful to the gardener in other ways. They provide a required background or border, hide an old stone wall or fence or even a compost pile, act as a protective windbreak for delicate neighboring plants and shelter for shade-loving ones. They also add a unique charm to your garden, with their bright colors ranging from white to yellow to assorted pinks to red and even purple. They are unequalled as bee attractors too. And don't forget to break up the second-year stems into 1- or 2-inch pieces and use them as extra mulch or compost material.

Honeysuckle

Honeysuckle
Lonicera caprifolium
Goat's Leaf

Learning by observation is good but learning by experience is even better, my grandfather used to say in one of his daily (and never-ending) sermons. And so, a long time ago, the honeysuckle shrub was given its synonym, goat's leaf, from the Latin words for "goat" and "leaf," because people discovered from daily experience that it was a favorite food of goats. In my travels along the eastern coast, I have found that many countryfolk today use the flowers, berries, and leaves for various purposes. Perhaps you will too.

For a mild diruetic to treat minor kidney problems, steep a heaping teaspoon of the flowers and ⅓ teaspoon of the seeds in a cup of hot water. Drink the liquid 3 to 4 times a day. Add honey or sugar to make a cough syrup and take a tablespoon every 2 hours. For kidney and liver complaints, decoct 3 or 4 teaspoons of the leaves in 2 cups of hot water down to one. Drink the strained liquid morning, afternoon, and evening. The unsweetened liquid is a good gargle base for a sore throat. And the dried leaves gathered in late fall come in handy as a skin-healing remedy for cuts, wounds, and sores. Boil a cupful of leaves in 3 cups of hot water down to one in a covered pot.

In Elizabethan times a conserve (concentrated syrup) of the flowers was fittingly kept in almost every gentlewoman's house. It's fun to make it today in your own kitchen. Gather the flowers at midday when there's no dew left. Prepare alternating layers of the flowers and raw sugar until the pot is full up. Then cover with herb-flavored vinegar (e.g., thyme or marjoram). Or you may "bruise" (cut or reduce to smaller size) well any amount of the flowers in a mortar or bowl, and add sugar boiled with water to make a thick syrup. Mix well. In either case add a small amount of lavender, thyme, or rosemary, or other fragrant herb. In Elizabethan times to "conserve" the flowers or other parts of the plant meant to keep them a year or two. The conserve was said to help asthmatic/bronchial discomforts, relieve cramps, and stimulate the kidneys. Use the syrup for the same purposes. And, too, it goes well with hot tea or may be mixed with jelly and taken with a cracker as a fragrant snack.

Hound's-Tongue
Cynoglossum officinale

How do you think the old-time herbalists, without research laboratories, microscopes, and other modern equipment, discovered this plant's therapeutics? Bite or chew a young leaf (an older one is too bristly), and then note that the texture is rather "distinctly" mucilaginous, which represented, claimed these astute observers of nature, the phlegm or expectorated mucus of the body. Externally, it signified the oozing pus of sores.

If you have need for a demulcent and mildly sedative cough syrup, steep a teaspoon of the leaves and root (both collected and dried in spring and early summer) in a cup of hot water for 10 minutes and allow to cool. (The teaspoon of herbs may also

include cardinal, flax, scabious, Jacob's ladder, or hollyhock.)
Strain and add only enough honey to prepare a thin syrup. Sip 2
teaspoons slowly every 3 to 4 hours. The syrup serves well in
bronchial catarrh and pulmonary complaints.

You'll find that the following decoction is a worthy astringent for skin cuts, insect bites, and ulcerated sores. Boil a heaping tablespoon in a pint of hot water for 15 to 20 minutes, and
allow to cool. Strain and apply the liquid as frequently as
needed.

Use tablespoonful doses to check diarrhea and summer
complaints, and the decoction as an injection for hemorrhoids.
(Both syrup and decoction should be refrigerated when not in
use.)

Make a healing ointment by stirring a heaping tablespoon
of the dried, ground leaves in a cupful of melted, salt-free lard
(or equal parts of lard and lanolin) for 15 to 20 minutes. It's
good for the above skin problems and for burns and scalds. (The
tablespoon of herbs may also include loosestrife, American ivy,
spruce, scabious, or periwinkle.)

Houseleek
Sempervivum tectorum

Here's a cooling drink for summer's heat, one that's sure to re-
fresh you and be chock-full of vitamins and minerals. Process 2
large cupfuls of this plant's washed leaves, 2 stalks of celery,
and 2 carrots through your juicer. Mix ¼ cup of the juice with
½ cup cold water and sip every 2 to 3 hours as required. "A
posset [a warm sweet drink] made of the juice," said
nineteenth-century British herbalist Nicholas Culpeper, "is sin-
gularly good in all hot agues [fevers], for it cooleth and tem-
pereth the blood and spirits." In the absence of a juicer, use the
blender and thoroughly squeeze out every drop. Take 2 tablespoons mixed with ½ to ⅔ cups of water as indicated. Refrigerate the balance.

The ancient herbal writers suggested other medicinal uses

for this plant. Galen recommended it in the treatment of shingles; Discorides used it for weak and inflamed eyes; Pliny, to produce sleep. Gerard stated that the juice was an active ingredient in the surgeon's populeon, a healing ointment.

Houseleek is a native of Greece, where, in ancient times, it was much used as a love potion and stimulating tonic. Perhaps this singular idea arose from the Greek's observation of the amazing benefits of eating the leaves, benefits that can be obtained today too. Overheated? Dehydrated? Exhausted? Try eating these fleshy, succulent leaves. They're a veritable reservoir of moisture, and they'll restore your vitality during the hot weather.

Include the short juicy leaves in fresh salads, with steamed or cooked vegetables, chopped with buckwheat groats or millet, in soups, stews, and other cooked dishes. Before using them in any dish, try soaking them for an hour or so in herb-aromatized wines or vinegars to make them take on a new flavor.

The expressed juice of the houseleek can be used in many remedies. It makes a cooling diuretic that is strongly indicated in feverish conditions, gout, and urinary problems. Simmer 2

Houseleek

heaping teaspoons of the finely cut leaves in 2 cups of hot
water down to one. Drink a tablespoon tepid-cool every 2 to 3
hours. To use as a warm gargle, steep a teabag in the finished
decoction and use it every ½ to 1 hour.

For scalds and burns, apply the crushed or mashed leaves as
a poultice; to alleviate skin inflammations, mix the expressed
juice in a little cold cream or with an equal amount of aloe
juice (which see), or mix both in a thin jell of quince and apply.
The same juice, or the leaf sliced in two, may be applied to
warts, skin eruptions, scratches, and fissures, and even mild
cases of athlete's foot.

Hydrangea
H. arborescens
Sevenbarks

The plant's common name means "water vessel," from the
Greek *hydr* ("water") and *angea* ("vessel"). In nature it is found
along banks of streams, rocky rivers, marshes, and other wet
places. Indian herbalists noticed the stem bark's tendency to
peel in layer after layer of varying colors and hence called it
sevenbarks. The Cherokees used a decoction of the plant in
treating calculous diseases (gravelly deposits).

When you need an active but gentle diuretic or antilithic
(an agent that prevents the formation of kidney stones and
gallstones), dry and granulate the summertime leaves, bark, and
twigs of this shrub and use a teaspoon of them along with a
teaspoon of an aromatic (lavender, New Jersey tea, or dictamnus) steeped in a cup of hot water. Drink tepid-cool. You'll
greatly reduce the inflammation of calculous (stone or gravel)
deposits, gradually effect their elimination, and relieve the pain
consequent upon their emission.

Want a spring, summer, or winter tonic or plain, old-
fashioned bitters for a run-down or rheumatic condition? Boil a
small handful each of barberry twigs, dandelion and American
ivy roots, and bachelor's buttons, all dried and cut, in a quart of

boiling water until reduced to half. Strain, sweeten with honey or raw sugar, and take 2 tablespoons in ½ cup of water 3 or 4 times a day.

For simple herb tea or as a substitute for Pekoe tea, dry the early summer leaves and steep ½ teaspoon of them plus the same amount of leaves or flowers, safflower, pansy, and bee balm in a cup of hot water. Keep covered 10 minutes and strain. Sip slowly and enjoy.

Drinking such health teas even once or twice daily should ensure your intake of rutin, since it is present in large amounts in hydrangea. Rutin helps to maintain the normal state of blood vessels by decreasing capillary fragility, common in hypertension. Its action is similar to vitamin P, for it too is bioflavonoid and comparable to the hesperidin (the white spongy portion) of citrus fruits. Rutin is also present in garden rue *(Ruta)*, pansy, forsythia, eucalyptus, and buckwheat.

I

Iris
I. florentina, I. germanica, I. pallida

Making sachets? The fixative you need to help retain the aroma of the pleasantly scented ingredients is the root of any of these species of iris. Gather the roots of 3-year-old plants, remove their corky outer layer, and dry them in the sun and later in a warm room. Have no fear as to their immediate acridity and absence of odor; during the drying process the acrid taste is lost, the distinctive fragrance is developed. Be sure to protect them against insect attack by placing a mothball on the base of an airtight container.

Ingredients for a sachet include whole rosebuds, violets, lavender flowers, and such cooking savories as thyme, rosemary, and marjoram, cloves and cinnamon (sticks), garden favorites such as lilacs, lily of the valley, and heliotrope, and especially

Iris

the rinds of lemon, orange, and tangerine. Make a 35 to 50 percent base of the rosebuds and lavender and add equal parts of the other ingredients to complete the mixture. Pour into the waiting mouths of 3 by 4 or 4 by 6 cloth bags. (Use discarded shirts, bed sheets, or pillowcases.) Prepare other — your own — mixtures according to recipes found in other sources.*

Bath Powder

Mix equal parts of powdered iris (orris) root, lavender, mint, bee balm, rose (flowers), rosemary, and/or pelargonium, and 2 parts powdered oatmeal. Put the mixture into a cloth bag, tie the opening with string, and either place in the bath or massage the body with it. Better than a brush and ever so much more delightfully soft and exhilarating.

Put orris root to further service: Powder it with a coffee or spice grinder, then sift it. It becomes your sachet powder to "sweeten" laundry instead of chemical products. Dusting it in the last rinse water adds a fragrant scent to your clothes as does sprinkling it over drying laundered articles and between ironed linens. Do place a ½-inch piece of the whole root between just ironed or still warm articles for extra fragrance.

A mixture of equal parts of powdered clove and nutmeg (and/or other sachet ingredients) and 3 or 4 times as much of orris will perfume your gloves.

Mix well equal parts of orris powder and sodium bicarbonate (baking soda), and you've made as good a dentifrice as any.

In my pharmacy I used to sell the long thick root as a teether for children and to be chewed by adults to mask or overcome an objectionable foul mouth odor.

* See also my *Better Health with Culinary Herbs*, Barre, Mass.: Barre Publishers, 1971.

American Ivy
Parthenocissus quinquefolia
Virginia Creeper, Woodbine

You've seen this prolific grower jump — not creep — up, up, up the stone walls of churches and college buildings. (Indeed a synonym, Virginia creeper, belies its rapid spread.) Let's make use of this extensive climber and its shooting growth of well over 15 feet each year.

Should your present need be a ground cover for a barren or sandy area on the outskirts of, or in your garden, just plant several 2-foot lengths generously over that spot. Let the spreading vine add immeasurable beauty to your unadorned fence and large trees.

If you know this plant as woodbine or wood vine, and correctly spell the latter syllable *vein,* you'll understand that since the lengthy, extending stems were thought to represent the bloodstream, they have been employed in medicine as a sarsaparillalike alterative in blood and skin problems. Mix equal parts of the dried stems, black birch, dandelion root, barberry, bittersweet twigs, and violets. Boil 2 heaping tablespoons in 1½ quarts of hot water down to half the quantity. Let cool, strain, and drink cold ½ cupful 4 times a day. Sweeten with a little honey. Especially good for teen-agers with acne, boils, or pimples.

To make a jiffy diaphoretic/expectorant for a minor bronchial discomfort, simmer a tablespoon each of hollyhock, violet, ivy leaves, thyme, and spiderwort in 1½ pints of water for 20 minutes in a covered pot. When cool, strain and add enough sugar or honey to syrup the liquid. Sip slowly a tablespoon every hour or two as needed.

We may not have an epidemic as horrendous as London's Great Plague, but we should recognize the role that ivy berries played during that time. Should a member of your family experience a lasting feverish cold or discomfort, use equal parts of the fruits and the four herbs mentioned in the previous paragraph to make a hot decoction (a tablespoon boiled in a pint of water for 15 minutes). Just as it did a century ago, this potion will produce internal and external perspiration, which lowers the troublesome fever and dispels the toxins that caused that condition.

English Ivy
Hedera helix

Use this familiar, hardy evergreen creeper to cover those barren parts of your shady garden grounds. Gather cuttings in late summer, water amply, and when they've developed well, place them in those spots sheltered from the sun.

Let the leaves' presence remind you of ivy's external application. A scratch, sunburn, beesting, running sore, or an old ulcer? Squeeze out the juice from the plant and apply several times during the day. English pharmacists prepare an ointment that reduces corns and callouses by incorporating the extracts of the ivy's parts into a lanolin-based cream. Prepare your own (and use it for other external purposes noted above) by heating a cupful of cut leaves and 1 or 2 teaspoons of the bark's gummy exudation in 1½ cups of Vaseline and lanolin (equal parts). Cook the mixture for 15 to 20 minutes at a low heat, stirring it occasionally, and strain.

A Commendable Remedy for Sore, Tired Feet
A soothing footbath quickly readied by vigorously boiling 3 large cupfuls of the cut leaves in 1½ quarts of hot water for 10 minutes and straining the liquid into a suitable pan. Soak the feet in the warm water, and equally important, massage the affected parts.

A Filling to Seal Decayed Teeth
Cut the bark (in the summertime), and a brown resinous substance will exude from the incision. Roll it between the fingers until toothpick-thin, and having previously dried out the cavity

with a Q-tip, insert and press in the gum gently with a finger. British chemists have sold the toothache remedy as "Ivy Gum."

Hail, oh Bacchus, god of wine and conviviality, the ivy is dedicated unto you. Years ago, an ivy bush, or a hand-painted facsimile, was placed on, over, or beside the door of English taverns and roadhouses to indicate that wine was sold there. Traditionally — and erroneously — Europeans have used cups of thick ivywood and the berries to supposedly dissipate the alcoholic content and the effects (drunkenness) of wine, leaving mostly water in the cup. The wood is intensely porous and absorbs only the wine's color in its passages. Mistakenly or not, to heal a liquor-hurt spleen (and presumably liver), a warm aqueous infusion was allowed to stand in an ivy cup for an hour and such a cupful was taken several times a day.

J

Jacob's Ladder
Polemonium caeruleum
Greek Valerian

Gather and dry the overground portion of this garden plant in the summertime. Use it as a hot tea (a heaping teaspoon to a cup of boiling water) to break up a feverish cold. To prepare an expectorant for a catarrhal cough and minor bronchial discomfort, add enough raw sugar or honey to make a syrup of the cool, strained infusion.

If you're encountering a mild or temporary case of fatigue, sleeplessness, or nervousness, drink a tepid-warm tea of the dried leaves and flowers as prepared above. It's best to drink the tisane on an empty stomach, i.e., after not having eaten for 2 to 3 hours, and to avoid all stimulating foods.

A Health Drink (or Tisane)
Of equal parts of its dried parts, leaves of violets, safflower, ground lemon peel, and mint, steep a teaspoonful in a cup of hot water (covered) for 15 minutes, stir, strain, and sip slowly every 3 to 4 hours as desired. If needed, add only enough honey to sweeten.

Call this plant Greek valerian if you will. English friends have told me that their cats love the smell of it and enjoy rolling on it, as they do on valerian (which see) and catnip. The only bond between the two valerians, however, is the shape of the leaves.

L

Lady's Slipper
Cypripedium parviflorum pubescens
American Valerian

Indian medicine men and herb doctors used medications made
from the roots of this plant long before the arrival of white
settlers. These learned practitioners employed the roots as a
fast-acting remedy in nervous disorders (thus another synonym
nerveroot) and female troubles of all kinds. (Note that *Cyp-*
ripedium is from *cypr* ["Venus"], woman or female, and
pedium, "foot" or "slipper." Moccasin plant, another synonym
for lady's slipper, is a corruption of the American Indian's
ma'kasin.) The Indians long dispensed assorted preparations for
the very same ailments for which the medical profession prescribed the herb as a fluid extract or tincture until only a few
years ago. White medicine men had considered it a safe, gentle

Lady's Slipper

nerve tonic and antispasmodic, and a worthy substitute for valerian. Perhaps its extract should replace chemical tranquilizers, which have dangerous side effects.

Your garden beds will offer an abundance of the ladies, and you may gather the fibrous roots (in September) for healing purposes. But if you or some adult in your family is experiencing a nervous headache, irritability, insomnia — whether or not caused by functional (female) or uterine discomforts — and general nervous conditions due to fatigue, then consider taking warm teas of the roots mixed with other herbs.

At one time, I made a nervine preparation that was composed of equal parts of mint, valerian, catnip, chamomile (or balm), and lady's slipper (roots). In the absence of one or two ingredients, substitute rosemary or motherwort. A warm infusion (a teaspoon to a cup of hot water) was (and can be) drunk 3 or 4 times a day to very good advantage.

Lavender
Lavandula vera

Over six decades have passed since that day when I first became acquainted with this herb. Grandfather and I were walking by a fruit stall when we came upon a cherubic elderly lady selling small bunches of fresh lavender flowers. "Place a sprig," she admonished us, "near the windows of your bedroom and you'll never fear the terrible epidemic of World War I." Later, at home, Grandfather did place sprigs of the lady's lavender with cloves of garlic (just for extra protection) in every room of the house. This feat, he stated some years later during one of his herb "sermons" to us Harris children, surely must have warded off the danger of infection. (Plus the fact that *all* members of our family were required to drink bitter "blood-purifying" herb teas.)

Gather the upper ⅔ portion of the plants before the flowers begin to bloom, during midday of the second successive July-August sunshiny day. Dry them well by suspension or on window screens. Separate the whorls of purple flowers from their terminal spike and the narrow grayish green leaves, and save the remaining stalks for an herb wine or vinegar.

For thousands of years, this perennial ornamental has been used in countless ways because of its unfailing, delectable fragrance. Sprinkle a few of the fresh flowers over your vegetable salad and soups. When preparing apple or other fruit jelly, set aside several jars for lavender's exotic flavoring. Before pouring the cooked juice, place a few of the flowers at the base of the glass, pour, cover with melted paraffin, and seal. Serve with desserts and tea.

In Elizabethan days, a conserve or syrup of lavender was used to "comfort the stomach," i.e., prevent possible stomach disorders, and was served with heavy meals as a jelly or preserve.

You may lavender-season your white port wine and use to prepare desserts, fruit gelatins, sauce, or jelly molds. Shake well 1 teaspoon of fresh (or dried) flowers and stalks in a cupful of slightly warmed wine contained in a sealed bottle and let stand for 2 to 3 days in a warm place. Stir daily and use as is or strain before using. The sauce in particular goes well with chicken, veal, or lamb. (You may also combine lavender with marjoram, thyme, or savory.)

To aromatize vinegar, combine a cup of the ground remaining stalks and leaves (and ½ cup of dillseeds, mint, and citrus peels), with a pint of warmed cider or malt vinegar. Let stand in a warm place for 2 weeks, shaking it every 2 to 3 days. Strain. Use it in your salad dressing, sauce, and especially as a warming marinade or baste for fowl, meats, or fish. The heat dissipates the acidity, leaving a refreshing herbal taste.

The compound vinegar serves as a gargle for a sore or irritated throat: Mix 1 ounce each of the vinegar and honey and a tablespoon of lemon juice. Gargle, using the solution as warm as possible, every 1 to 2 hours. The vinegar, mixed equally with witch hazel extract, becomes your speedy remedy for insect bites. Saturate a cotton ball with the liquid and apply every ½ hour.

An Insect Repellent

Digest in a pint of ethyl rubbing alcohol, previously warmed, a cup of equal parts of pennyroyal, tansy (flowers), mint, wormwood, and lavender, all dried. Let remain for one week and shake the bottle once daily.

For a quick-acting, penetrating liniment for a charley horse, painful bruise, or swelling, mix equal parts of the above vinegar, witch hazel, and turpentine, and for every 4 ounces, add ½ teaspoon each of ground red pepper, ginger, and/or cloves. Warm the mixture, strain, and apply as warm as possible. (During World War I, lavender's freshly extracted oil was used by the Allies to treat open wounds and later studied in full by the French Academy of Medicine for other various antiseptic applications.)

My Aunts Mary and Esther knew lavender's charm well. Before their beaus made an appearance, the ladies shampooed their hair and followed with a rinse of strongly aromatic lavender flowers. You might try it. Place 1 or 2 teaspoons of the flowers in a pot and slowly add a pint of hot water, stirring. Cover for 20 minutes or until tepid. Strain into a basin and rinse the hair. The ladies doubtlessly used another kind of post-shampoo rinse of lavender, rosemary, and sage to add extra sheen to their black hair.

I daresay that my wily aunts also placed sprigs of the plant here and there, just as in past centuries the flowers were strewn on the floors of churches and homes on days of festivities or celebration. They also may have kept concealed several open jars of the lavendered vinegar behind the sofa and chairs, on the piano and knickknack shelves, and other inconspicuous spots in the parlor.

Interestingly, our commercial toilet waters first began with vinegars of lavender and other favorite aromatics. Here's an easily prepared formula: Mix thoroughly in a pint of malt or cider vinegar ½ ounce each of lavender flowers, mint, dried lemon or tangerine peel, rosemary, carnations, and a pinch of cloves. Let stand corked near warmth or in the sunlight for 10 to 15 days, stirring once daily, and strain. Replace any ingredients with or add small amounts of orange peel, red bergamot, basil, marjoram, mints, and other spices and/or herbs of your choice. Too strong? Dilute the solution with 1 or 2 ounces of water. This is one of the products prepared (with rubbing alcohol as a base) by the students of my herb-study class.

Note that the generic name for lavender is derived from the

Latin *lavare*, to wash. It refers to the wealthy ancient Romans'
use of the flowers to perfume their bath and to the time-
honored custom of washing linens with lavender. Thus, since
the twelfth century, a washerwoman was called a lavenderess,
hence laundress. A good reminder when you set your linens in
the closet: Put whole dried flowers between bed linens, and
with other clothes stored in coat closets, in a trunkful of
clothes, or in the attic. Helps to keep the moths away.

When spring cleaning, derive the same practical benefits (of
moth prevention), and the addition of a far more pleasing aroma
(than camphor balls), by virtue of an antimoth mixture of equal
parts lavender, rosemary, lemon peels, tansy, wormwood, and
rue. This mixture may further be augmented by such good old
reliables as thyme, santolina, and sweet flag root (obtainable in
pharmacies and health-food stores). Put a tablespoon in small
cloth bags and suspend them in clothes closets and place them
in drawers and pockets of stored coats.

When preparing a potpourri, include the same sachet ingredients. Add flowers of peony, pansy, hollyhock, and phlox, food
seasoners such as basil and cardamom, and especially a few
drops of such aromatic oils as eucalyptus, lemon balm, neroli,
sassafras, verbena, and others. Use orris root as the fixative, ½
ounce to every pint of rosebuds (or petals). The herb mixture
should set for 2 to 3 weeks packed into small glass jars. Potpourris make nice gifts.

The compound tincture of lavender I sold in my pharmacy
was really a variation of a centuries-old remedy for a most valuable nervous tonic and cordial for indigestion. Its components
were lavender flowers, rosemary leaves, cinnamon, nutmeg, and
several other aromatics (with or without red sandalwood). They
were soaked in *strong* wine (i.e., with a high alcoholic content)
for 10 to 15 days. The strained liquid was taken in teaspoonful
doses 4 times a day.

Until recently, an infusion of the flowers or a syrup (i.e., a
conserve) of the young tops just flowering was prescribed by
herbalists and physicians for nervous headache, vertigo, and
palsy. English herbalists recommend the tea to lessen nervous
tension and headache due to fatigue, both as a drink and as a
wet application over the forehead and temples. For everyday
use, make a tea by stirring ½ to 1 teaspoon in a covered cup of

hot water for 7 to 10 minutes and strain. Drink 2 tablespoons warm-tepid every 2 to 3 hours as needed, *slowly,* and relax and enjoy.

The infusion is also good for flatulence or nausea, in which case take a tablespoon every hour if necessary. It is a good carminative, mild sedative, and stomach tonic. (A stronger tea proved quite effective in World War I for swabbing infected wounds.)

Lavender Cotton
Santolina chamaecyparissus

This plant is related to neither lavender nor cotton. A quick rub of its distinctive foliage reveals a clean but disagreeable scent, very different from that of lavender.

Yet, as Mesdames Marjorie and Roberta discovered, this attractive edging can dress the borders of gardens and, unfortunately, carpet the adjacent areas as well when not kept within bounds. The attractive silvery gray leaves and the small buttons of yellow flowers can form the background for fresh (and dried) floral arrangements.

Discover, as they and other students of my herb-study classes have, that cabbage plants can be free of the harmful insects that usually attack members of the cabbage family. Correct! Lavender cotton is there to repel them.

Let lavender cotton also render its moth-chasing and clothes-protecting services. Dry and finely grind the foliage and flowers and mix a portion equally with (true) lavender, arborvitae, feverfew, and/or pyrethrum. Stir well for even distribution and spread a tablespoon of the mixture in cloth or nylon material. Close all sides and place one such "package" between your woolens, in each corner of all usually closed clothes closets, and in the lower drawer of your dresser. Sprinkle very lightly under the corners of your carpets too.

To further antimoth your attic, place a bouquet of the aforementioned plants in the open sun of each window. And place one of the packages among the clothes, blankets, and other such items stored there.

Grandfather, that imaginative herbalist, would hang such aromatics around the exposed electric light bulb on the un-screened porch so that folks could enjoy the balmy summer weather and not be pestered by mosquitos.

Give your dog a break too. Help free him of annoying pests by preparing his very own (anti-) flea powder. Grind finely 2 parts of the dried flowers and leaves and the ingredients of the previously mentioned antimoth mixture and sift. With some force, rub the powder into the hair as near to the skin as possible, going from the tail to the head. And be sure to sprinkle some of the powder in the dog's box or kennel or wherever he lodges.

Lilac
Syringa vulgaris

Let's put the exciting, exotic aroma of this shrub's flowers to better service. They appear in early spring, and the fragrance they exude signals they're ready for action. Be sure to include them in milady's sachet. Collect stemfuls of the *unopened* flower buds and dry them while still on the stems and *away from the sun.* Yes, Menasha, writer-student of my herb class, you *should* remove the floral parts but do save and chop the stems for your herb tea.

You may prepare different varieties of sachets by mixing in various proportions the ingredients mentioned under lavender, violet, and rose (which see). Be sure to include any of your favorite food-seasoning herbs and spices found in the kitchen cupboard. Potpourris and herb pillows also ask for lilac. And perhaps you'd like to try my health tea (a tisane), made by steeping any mixture of vegetal ingredients (not spices) of the above products. Of the teaspoon to be steeped in a cup of hot water, use ½ teaspoon of lilac flower buds.

Lilac

Remember them too when you're about to make fruit preserves, jams, and especially jellies for meat and poultry. And
try them as well when you're aromatizing wines and vinegars.

Remember that your lilac bush will help you to overcome
a cold or fever. Gather the leaves and flowers when the latter
have fully bloomed, dry them *well,* and grind the leaves to infusion size. Drink a warm tea (heaping teaspoon to a cup of hot
water) every 2 hours. (It is best to eat as sparingly as possible.)
Perhaps your grandparents still recall that during World War I's
"great epidemic," the lilac flower-leaf infusion was much used
as a fever chaser in influenza and malaria. Recuperating from a
recent illness? Consider the cold infusion as an invigorating tonic. Sweeten with the smallest addition of honey.

Lily of the Valley
Convallaria majalis

The second Latin name of this plant, *majalis,* means "belongs to May," and that's when to gather the white, bell-shaped flowers. Only be careful! You'll be competing with the bees who are after the pollen in the flowers.

Let their sweet essence, formerly the true scent of muguet-named perfumes and other toiletries, suggest that you prepare your own wrist posy, nosegay, or corsage. Cut them at the base, place in water until needed, and then tie them with a suitably colored ribbon.

Do include the freshly dried blossoms in your sachets (called "sweet bags" in olden times), in company with lilac, lavender, and violet (which see), and a quarter to half the quantity of the ever available dried citrus peels.

A Dyeing Experiment
Boil the leaves, strain, add a little alum or ammonia water and get a yellow shade; or add limewater for green.

Lily of the Valley

When, in May or June, you're considering making homemade wine, don't neglect the pearly flowers as another floral ingredient.

Capture the delicate bouquet of the lilies by stirring a cup of the fresh flowers in a cup of vegetable oil and placing the capped container either in the sun or on the warm oil burner or kitchen stove. Let this simmer for 2 days and gently shake or stir twice a day. If the flowers are to impart their sweetness to the oil, the procedure must be repeated at least 10 to 12 times. It's worth it.

A Cultivation Experiment

You can hasten the growth of undeveloped lily plants by using an anesthetic such as chloroform (obtainable at pharmacies). Wet a cotton ball with a few drops of chloroform and let the winter buds absorb the vapors for a few hours. Then plant the buds. They'll invariably develop leaves and flowers far sooner than others not so treated. The result is surprisingly good.

Linden
Tilia americana, T. europaea
Lime or Basswood

Tiliae ad mille usus petendae: "Lime trees [are] demanded for a thousand uses." The synonym lime tree applies to both species, though more specifically to the latter, and foretells their usefulness as an antacid or systemic alkalizer. The flowers and leaves are used in medicines. Their mild diaphoretic property is indicated to eliminate the undesirable catarrh from the bronchia and stomach. John Gerard recommended that the flowers be used "against paine of the haid proceeding a cold cause, against dissinesse, the apoplexie and also falling sicknesse, and not only the floures but the distilled water thereof."

Enjoy the benefits of your small-sized linden tree or the

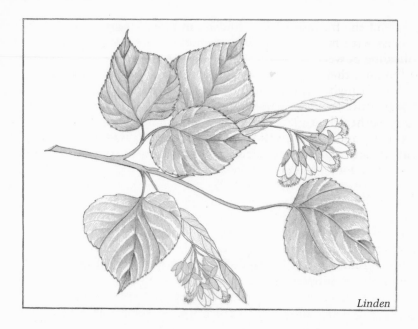

Linden

huge basswood on your property. Perhaps you already know that lindens are a favorite in European cities. They are planted along avenues where folks promenade, rest, and enjoy a cup of tea made with the infused linden leaves and flowers. Some people are revived simply by its delicious scent.

To make tea, gather and dry the leaves and flowers. Steep a heaping teaspoon of these parts in a cup of hot water for 10 minutes, strain, and sip slowly.

Powder the parts in a coffee or spice grinder and use with powdered aromatics as a salt substitute.

Professor Heber W. Youngken, of the Massachusetts College of Pharmacy, considered the yellowish white flowers and light charcoal of the wood excellent and quick-acting remedies for dyspepsia and other digestive disturbances. The charcoal is prepared by burning a few fall-gathered twigs in your fireplace, scraping off the charred effects, powdering and stuffing it into empty gelatin capsules, which can be purchased at your drugstore. A capsule is taken every 2 to 3 hours (or as needed) and followed with a tepid cup of aromatic tea, made of either ingredient.

In my pharmacy I always enjoyed a brisk and continued

sale of the fragrant linden flowers, and I had lengthy discussions with my French and German customers regarding their healing powers. To them, Indian or Chinese tea was a supreme "no-no"; their choice, of course, was *tilleul* or *linde*. It was (and is) their nightcap and ours to adopt. (See above for directions.) Substitute the flowers (and leaves) for Pekoe tea and you will have a health drink as well as a simple remedy for cramps, spasms, indigestion, and other gastric complaints. You'll find the tisane a pleasantly flavored and soothing antispasmodic and sedative for sleeplessness and various nervous situations. For best results, use only the flowers and leaves of the present and immediately past summer's collection.

A Catarrh Remover

Simmer ½ cupful containing equal parts of the leaves, hibiscus, and veronica in 1½ cups of hot water in a covered pot for 15 minutes. Strain and, when cool, add enough brown sugar or honey to make a syrup. Take a tablespoon every 2 hours as required. It's also indicated in colds and bronchial catarrh.

Try This at Winter's Close or
Early in the Season

Draw the sap—and a considerable amount of a highly nourishing sugar—from the larger basswood. Use it as you would maple syrup.

A Hand Lotion

Simmer a cupful of the mucilaginous leaves and shoots in a pint of hot water down to half the content. Let cool, strain, and apply cold to cuts, sores, burns, chapped hands, and to relieve minor skin problems.

Live-Forever
Sedum telephium
Orpine

Bite into the fresh succulent leaf and you will learn to enjoy the bittersweet and faintly astringent taste of this plant as well as its juicy and cooling effect. A good food and medicine!

Wash the delicious spring leaves and stems and mix them with other salad vegetables. If you gather them through June, steam them with vegetables or add them to soup or stew. Make a relish by cutting them into quarters and pickling them in a herb vinegar for several days. This makes quite a tasty condiment for fish or meat; chopped even more, they may be included in a coleslaw.

If you have a superabundance of the plants, remove a mass of the crisp tuberous roots, brush/wash, and either cook them a few minutes before you eat them or pickle them in an aromatic vinegar a week or so.

The succulent leaves will provide you with a quick remedy for skin irritations, insect bites, pus-forming sores, and slowly healing ulcers. Crush them completely and apply. To alleviate painful fresh burns or scalds, boil the cut leaves in hot milk (2 heaping tablespoons to a cup of milk) for 5 to 6 minutes, let cool completely, and gently sop the cold liquid onto the affected area. And let the strained but still warm decoction be your medicine for diarrhea or loose bowels. Take in ½ cupful amounts 3 or 4 times a day.

For a temporary kidney disease, steep a cupful of the cut leaves in a pint of hot water for 10 to 15 minutes, allow to cool, strain, and drink a cupful every 2 or 3 hours. To facilitate the healing process, eat only sweet fruits, e.g., apples, grapes, melon, during the day.

Black Locust
Robinia pseudo-acacia
Yellow Locust, False Acacia

Gather the pendulous clusters of these creamy white, fragrant flowers and let them share equal space with lilacs in a wide-mouthed glass bowl of water. Their fragrance resembles the sweet pea's.

A Challenge
The American Indians cooked the oily but slightly acid seeds in water to dissipate their acidity. The water was removed and the seeds saved to be used as an easily digested, wholesome food. You may do the same today. For future winter use, gather the seedpods and label the container.

In my youth, folks would include the seeds in soups, stews, or vegetables, without knowing their high content of fat, carbohydrates, and protein. The oily seeds were also roasted, often with squash or pumpkinseeds, on top of our hot kitchen stove and eaten warm with the evening meal or as a snack. M-m-m, I can still savor its sweet, chestnutlike flavor!

Perhaps you have need of preventing soil erosion along the slopes of your property. Plant rows of these small-sized shrubs. The spines of the larger locusts increase their usefulness as a hedge.

Purple Loosestrife
Lythrum salicaria
Purple Willow Herb, Red Sally

Gather and dry these attractive plants as they begin to blossom. Drink a cup of tepid-warm infusion for a minor liver, gallbladder, or blood problem. (A teaspoon of the ground herb to a cup of hot water.) For a feverish cold, drink the tea warm-hot 3 to 4 times every hour.

Keep on hand a quart-sized container of the (dried) fall-gathered parts — the leaves and flowering tops — and remember that an effective corrective for simple diarrhea can be made from them because of their demulcent/astringent qualities. For this purpose, boil a teaspoon each of these parts and cranesbill, ½ teaspoon of sage and bugle in a pint of hot water in a covered pot down to half the content. Strain, stir in a large pinch (or ⅛ teaspoon) of powdered cloves and cinnamon, and take a tablespoon warm every hour. It is best not to eat or drink until the condition is corrected.

Also use this warm liquid as a gargle and mouth rinse for sore throat, quinsy, and mouth cankers.

For sore or irritated eyes, gather the purpling flowers in June, dry them, and steep a teaspoon in a cup of hot water up to the count of 10. Cover until cool. Strain through filter paper or several thicknesses of cloth and drop into the eyes every hour or so. May be used with barberry and rose (which see). Refrigerate the balance. The solution may be used as a cold compress with an equal part of witch hazel extract.

When the flowers are truly purple (August) they may be used to relieve liver and gastrointestinal discomforts. Gather and dry them and the leaves by suspension in a warm room or attic with good air circulation. Mix equal portions of this, hawkweed, hollyhock, and hepatica with a bare sprinkle of saffron stigmas. Drink hourly teas of the mixture (a teaspoon to a cup of hot water).

You may want to "be prepared" with a healing salve for sores, cuts, and indolent ulcers. Gather the roots and leaves in early fall, dry them well, and cut them into small pieces. Gently simmer for 15 to 20 minutes a heaping tablespoon in a cup of melted unsalted lard, and then add a tablespoon each of brown sugar and beeswax. Simmer another 5 minutes until incorporated and pour into a suitable jar.

White Lupine
Lupinus alba

In her *Modern Herbal*, Maud Grieve talks about an extraordinary lupine banquet held in Germany in 1917 that best exemplifies the plant's multifarious uses. To wit: The table was covered with a tablecloth of lupine. A soup of lupine was served first followed by a lupine steak roasted with lupine oil and seasoned with an extract of lupine; slices of 20 percent lupine bread with lupine margarine and lupine cheese were served; and the dinner ended with lupine liqueur and lupine coffee. Finally everyone washed their hands with lupine soap. To write home about the banquet, guests used lupine-fiber paper and envelopes with lupine glue.

Historian Pliny wrote that the Romans cultivated lupines as a plant food. He praised its wholesomeness and its easy digestion. "If taken commonly at meals," he said, "it will contribute a fresh color and a cheerful countenance." He believed that a diet of the seeds sharpened the mind and augmented one's imagination. In *Elixirs of Life,* Mrs. C. F. Leyel, Fellow of the Royal Institution of Great Britain, notes that "Lupines have an ancient reputation for increasing courage and sharpening vision. A celebrated painter of Rhodes called Protogenes lived on a diet of lupines for several years, while he was painting the famous hunting piece of Ialysus, the founder of Rome."

These notes on recent medical research with lupine concern us all: The sprouts have been employed as a new simple test for cancer and to determine the advance or withdrawal of the malignancy cells while under treatment; a constituent named magolan has been found useful in diabetes mellitus; a

White Lupine

decoction of the seeds had displayed a marked ability to increase sugar tolerance in diabetic patients, suggesting a practical replacement for insulin. Lupanine, another plant substance,
has strong antimalarial properties.

Here's a more down-to-earth use of lupines: You may use
the withering leguminous plants as green manure in fall. Scatter handfuls over the gardened area, or make them an ingredient of your soil-enriching liquid compound. While a teaspoon
of animal manure digests for 2 days in a jar containing 5 gallons of water set in the sun, gather a tablespoon each of
lupines, weedy characters like pigweed, dandelion, and burdock,
and a teaspoon of sage and chamomile, all to approximate a
cupful. Mix into the liquid and let stay in the warm sun another 2 to 3 days. Add a cupful to each 2 to 3 gallons of water
and spray or sprinkle over recently turned soil. Do the same in
mid-spring. Lupines possess a strong nitrogen-assimilating
power and produce a green manure for agriculturally
"worthless" — light and sandy or even barren — soil, where very
little of economic value will grow.

Use the seeds to make a facial and skin-softening face
mask. Use a spice or coffee grinder to reduce the seeds to a fine
powder and mix together a tablespoon of the seeds, uncooked
oatmeal, a teaspoon of honey and lemon juice, and 2 tablespoons of cucumber juice, and either add enough witch hazel
extract to soften the mass or incorporate the previous ingredients in a lotion of quince seeds (which see). Or merely mix
the powdered seeds and oatmeal in a base of avocado or but-
termilk. And incorporate the powder in water-thinned aloe
juice (which see). A strong decoction of the seeds (a tablespoon
to a pint of hot water boiled 30 minutes) may be applied cool-
tepid to the usual skin affections and scabby ulcers.

Should you ever hear or read about the Latin saying *mum-
mus lupinus*, remember that it means a spurious bit of money.
Roman actors used lupine leaves as money in their plays.

M

Magnolia
Magnolia

Know anyone that wants to stop smoking? Take a cue from the
herbalists' suggestion that the stem bark of the *Tripetala* variety
of magnolia is a sure cure for chronic tobaccoism. Tell the
chain smoker to chew a piece of the bark or twig as often as
(s)he would reach for a cigarette or cigar. I used to prepare such
an antismoking remedy with the powdered bark, which I
stuffed into #00 gelatin capsules, Two were taken with a pleas-
ingly aromatic tisane morning, midafternoon, and 1 or 2 hours
after supper.

   Drink a tepid tea of the infused (summer) bark (and/or
twigs) as a nonastringent tonic and sweetening bitters in dyspepsia and mild stomach disorders. Take it warm-hot to increase sweating in colds and feverish affections.

Magnolia

A dear herb-using friend and customer, octogenarian Mrs. J. Murphy, told me that in the "ould country," country folk considered the tree of great worth in chronic rheumatism, especially "when the weather changed." A large handful of the *early* twigs was allowed to steep on a hot stove for a day, and a cupful was removed, strained, and taken 3 or 4 times a day. It would gently move the bowels, cause an internal sweat, and therefore help to detoxify the system and slowly relieve the pain. It is worth a try.

Maidenhair Fern
Adiantum pedatum
American Maidenhair
A. Capillus-veneris
True or European Maidenhair, Rock Fern, Venus's-hair (Fern)

The generic name of this familiar household plant is from the Greek *adiantos*, "unwetted," and refers to its water-repellent property; the word was used by Theophrastus and Pliny to describe Venus emerging from her watery abode, the Mediterranean Sea. The plant has also contributed to hair remedies. In my early pharmacy days, I sold a Capillus Hair Tonic.

By observing that the thin shiny leaf stalks resembled a hardy growth of hair, the ancients determined that maidenhair fern would stimulate hair growth. For the past two hundred years, preparations of the plant have been well accepted as hair tonic in continental Europe where herbalists still manufacture the product.

Make your own hair/scalp invigorant. In a covered pot simmer a large handful of the dried leaves, a tablespoonful each of yarrow, rosemary, and wormwood, and ½ tablespoonful of sage in 2 pints of hot water. Cook until about a pint remains. Set aside and strain. Rinse the hair with cold water, dry off the excess, and apply the cool preparation to the hair *and* scalp with either all 10 fingertips or a natural bristle brush. Towel-

turban the hair until completely dry and brush the hair vigorously.

John Gerard, renowned Elizabethan herbalist, tells us that
"it maketh the haire of the head or beard to grow that is fallen
and pulled off." And the high praise that Nicholas Culpeper,
the magnificent seventeenth-century London herbalist-
physician, bestowed upon white maidenhair, *Asplenium
ruta-muraria,* applies as well to the Adiantum species: (it is)
"singular good to cleanse the head from scurf [scales or flakes]
and from dry and running sores [of the scalp], and causes it to
grow thick, fair and well-colored for which purpose boil it in
wine, putting some smallage [wild celery] seed thereto and af-
terwards some oil." In recent years, the ashes of the fern were
incorporated with olive oil and herb vinegar and applied locally
for alopecia.

Equally important, the whole plant — root, stalk, leaves,
and stems — has been used since antiquity as a demulcent and
expectorant in bronchial and catarrhal complaints. It quickly
relieves the symptoms of shortness of breath, hoarseness, and
coughs resulting from colds.

French pharmacists prepare a cough syrup popularly called
Sirop de Capillaire by simmering 5 ounces of the *dried* herb
plus 2 ounces of licorice root in 5 pints of hot water for 5 or 6
hours. Up to the 1920s it was sweetened with honey (and now
3 pounds sugar) and flavored with orange flowers (now orange
juice).

Cough Syrup

Boil 2 heaping tablespoonsful of all parts (dried) in a quart of
hot water for 15 minutes and simmer for one hour. Strain and
add a pound of raw brown sugar. Again gently boil until the
liquid thickens, stirring occasionally. Cover until cold and bottle. Slowly sip a tablespoonful every 2 hours or as needed.

Tonic

Steep 2 tablespoonsful of the dried, ground leaves (fronds) and
rootstalk and one tablespoon of barberry, American ivy, and
pelargonium in a pint of hot water for 20 to 25 minutes. Strain
and take cold 2 tablespoonsful between meals and 2 hours before bedtime. Sweeten with honey to suit taste.

Until recently, poor people in western England and Ireland made a tea substitute of the bitter aromatic leaves. The Irish indigent, says Edward Lewis Sturtevant,* ate the leaves as a cooked vegetable. In time of need, I presume, the sharp dandelionish taste of the demulcent leaves could become quite acceptable to every palate.

Interesting that the Cherokee medicine men knew of its fever-chasing and antirheumatic qualities. They reasoned that since the fronds of the young plant are first curled but later straighten out, so too the potions, taken often, help to relieve familiar pains in the joints and straighten out tightened muscles caused by rheumatism.

Maple
Acer

Sure you know about the nutritious maple syrup and other sugary products made from the tree's sap. So here's a simple way to obtain the sap with the least effort. In late February and March, hammer a large (5 to 6 inch) nail, or spike (open) spigot into the base of the tree and hang a pail or bucket at its end. (The Indians used to slash the bark and insert a reed trough.) Remove the sap to your kitchen, boil it down to ⅓ the liquid content, and freeze it, either as is or as cubes. Expose a pint amount to room temperature for 2 weeks, and you've got an organically pure vinegar with which to prepare a sauce, marinade, salad dressing, etc.

Be a truly enterprising conservationist/householder and make further uses of the tree. Use the dried, fallen, or purposely cut twigs and limbs for your fireplace and the resultant alkaline charcoal as medicine. For temporary stomach disturbances, powder and sift it, and tap it into the #00 empty gela-

*Notes on Edible Plants, edited by U. T. Hedrick. Albany: J. B. Lyon Company, 1919.

tin capsules. Take 2 with a tepid aromatic tisane 3 times a day and at bedtime. The large pieces of charcoal can be used as an artist's crayon.

Please start thinking about winterizing in autumn. Use leaves that have fallen to cover those garden plants susceptible to frost. Don't indulge in the wasteful practice of discarding bags and bags of the fallen leaves. By all means do brush them against the foundation of your home and receive extra heat protection for the winter. Or feed a tree by solidly packing its base with thick, heavy layers of the leaves, alternating with a layer of heavy or large-sized flat stones or wooden boards.

Feed yourself with the washed young leaves. Add them to salads, to steaming vegetables, to soups and stews. Chop, grind, or dice them, and blend them into your casseroles, homemade fish and chicken croquettes, coleslaw, and with other chopped vegetable leftovers, into a chop suey or rice-vegetable mixture. You'll enjoy the sweet taste and slightly jellylike effect of the leaves.

You know the dangers of using white table salt. Hence make a salt substitute. Powder the spring-summer leaves via a coffee or spice grinder, mix thoroughly with powdered kelp and aromatics like marjoram, basil, oregano, and others of your liking. Use with all cooked foods, even with toast and with boiled or fried eggs. Gather many of the seedlings in spring, fast-rinse them in cold water, and use as above indicated re the leaves. When heating a soup, use them whole or cut but add them toward the end of the cooking. In fall, do gather and dry much of the countless sprouting seedlings for similar use during the winter months.

A Nourishing Tisane
Steep ½ teaspoon each of the dried leaves, mint, and carnation in a cup of hot water, cover for 10 minutes, stir, strain, and sip slowly. M-m-m!

I and members of my herb-study classes have used a decoction of the dried leaves and pounded twigs and inner bark of the red maple as a soothing compress for sore eyes. Or you may use equal parts of that liquid and witch hazel, with or without a simmered solution of rose petals (which see).

Make your own black dye or ink: Gather the dusky red bark of the ornamental red maple and boil any amount in hot water to produce a purplish color that turns black when you add lead sulfate.

And maples make very good windbreakers.

Marigold
Calendula officinalis

Oh, Marigolds! You rank high among all garden plants. You're the easiest to grow, you require the least attention, your blooms last until the frost arrives, and you offer an optimum of benefits. That's my idea of a good and dependable rent-paying tenant of my limited garden space. Alas, my poor distressed wife: "Must you have all those marigolds on almost every win-

Marigold

dowsill during fall and winter?" Well, why not grow them every month of the year, since the plant's Latin name means "calendar."

Once you've enjoyed marigold in any of a hundred ways, you'll never again ignore their pleas to serve you nor will you ever again treat your golds as mere pretty-pretty do-nothings. You too will take a cue from rapidly increasing numbers of health- and herb-oriented folks who now use the flower heads as a food seasoner, tint-dye, hair rinse, cosmetic, healing remedy, and delectable drink.

Gather the blossoms (and leaves) when they're fully expanded and dry them indoors away from the sun.

They're an acceptable replacement for the expensive saffron (see Crocus) to season/tint rice, potato, other starchy foods, and especially Mexican- or Spanish-type dishes, the better to prevent the overstay of harmful catarrh. Use the ground florets in soups, stew, and fish chowder, and as a paprika substitute with meats and poultry. Add a thin sprinkling to your next rice or cheese dish or gelatin puddings (blancmange desserts), and to your breads, pastry, and other baked goods. (Mrs. Van S.P. flavors her summer custards and puddings exotically with the fresh florets.)

A few other tips for using marigolds:
• To impart both a subtle flavor and a golden color to a salad dressing, add a teaspoon of the dried petals (plus other aromatics) to each cup of cider vinegar, add the required amount of oil, place the container on a warm radiator or stove for 2 to 3 hours, let cool, and refrigerate for a few hours. Remove and when it is liquid again, shake and serve.
• To flavor scrambled eggs: Beat 2 eggs, 2 tablespoons milk, and 1 teaspoon cut chives or scallions. Add to the pan containing melted butter and, when nearly done, add ½ teaspoon of fresh chopped marigolds and a pinch of basil or marjoram. Serve on or with whole wheat toast. Use the fresh golds in an omelet and over poached eggs, and add a tiny bit of the powdered florets over soft- or hard-boiled eggs.
• To color/flavor cottage cheese, yogurt, and cream cheese, add enough warm water to a teaspoon of the florets to yield a mild yellow, and incorporate.

◆ Fish chowder. Add 3 or 4 petals near the end of cooking.

◆ Casserole and fricasseed chicken. As above.

◆ Irish moss jell. Boil ½ ounce of the moss in a pint of hot water for 20 to 30 minutes, stirring occasionally, and, when the cooking has stopped, stir in a 3-finger pinch of the ground florets, and cover for 10 minutes. Express the jell, add a teaspoon of honey to each cup, and chill before serving. An excellent food/tonic for convalescents.

◆ Homemade wines, e.g., elderberry, dandelion. To each gallon of water, add 2 pints of the fresh flowers as one of the ingredients and let ferment according to directions.

◆ A liqueur. Add a cup of mixed herbs — a large handful of marigold flower heads and smaller amounts of carnation, marjoram, chamomile, woodruff, dictamnus, lemon balm, ground orange peels, and cinnamon — to a quart of brandy and let digest 4 to 5 days. Strain. Prepare a syrup of raw brown sugar (1½ cups dissolved in enough water, boiled, and allowed to cool) and add to the brandy.

◆ Making an herb vinegar? Add the florets for extra color and fragrance.

◆ Use the golds to replace the expensive saffron.

◆ A hair tint. Herbalist William Turner (1568) wrote: "Some use it to make their heyre [hair] yellow wyth the floure of this herbe not beinge content with the natural colour which God hath gyven them." Still, vanity will be served. Simmer a cup of the large-sized, bright yellow orange flowers in 1½ pints of boiling water for 20 minutes. Strain and use as an after-shampoo rinse, rubbing vigorously or brushing. A vinegar of marigolds will also lighten the hair and bring out the blonde highlights.

◆ To dye wool, cotton, or silk yellow, golden yellow to deep orange, first prepare a strong decoction of the dried flowers. Boil 3 cups of the dried flowers in a quart of hot water for about an hour, in which dissolve the mordants, 2 teaspoons of alum, and 1 teaspoon of cream of tartar (or chrome). Then boil 4 ounces of material in the liquid for about an hour.

◆ Healing benefits. Have on hand a marigold vinegar for purposes other than a hair rinse and food seasoner. A three-hundred-year-old "soveraigne remedie" quoted in Eleanor Sinclair Rhode's *A Garden of Herbs* (Boston, 1936) for a throbbing toothache or gum infection consisted of soaking a flower head in vinegar and rubbing it well onto the gums and teeth. (I have often prepared a tincture of *Calendula* — an alcoholic solution of marigold flowers — for several dentists who applied it in localized infections and in postextraction situations.) Use the vinegar as indicated in the following paragraph.

◆ External applications. Macerate for 7 to 10 days 2 heaping ounces of *grain* rubbing alcohol. Strain, dilute with an equal amount of water, and use for cuts, bruises, minor skin affections, beestings, and insect bites.

◆ Prepare a cosmetic astringent or skin cleanser by covering the ground flowers with just enough witch hazel extract. Let steep a week, stirring it daily and strain. Or simmer *very gently* for 5 to 6 minutes a heaping teaspoon of the florets in about 4 ounces of cold cream.

◆ An all-purpose healing ointment: Slowly simmer for 10 minutes 2 heaping tablespoons of flowers and leaves in a cupful of unsalted lard. Strain through cheesecloth and store in the refrigerator.

If you're an organic gardener, you're well aware that the root exudates of various marigolds provide an easy way to control certain nematodes (microscopic worms) that attack potatoes, tomatoes, tobacco, strawberries, and various bulbs (e.g., narcissus). Interplanting roses with marigolds helps to restore strength to the roses damaged by the harmful nematodes. Marigolds are reported to be also helpful in inhibiting wireworms, bindweeds, and ground ivy, and in regenerating depleted soil.

You don't have to be ill or discomforted to partake of the goodness of marigolds. When Pekoe tea is to be served, add a large pinch of the petals, fresh or dried, and your "cup [will] runneth over" with healthful benefits. And to ready a lotion for tired or inflamed eyes, dilute the well-strained liquid with 5 parts of distilled water.

Remember the flowers in all gastrointestinal affections: stomach cramps, colitis, peptic ulcers, and diarrhea. In such cases, mix equal parts of gentian, pelargonium, bee balm, linden, and marigold, and drink a cupful of the infusion (a teaspoon to a cup of hot water) every 1 or 2 hours. Add turtlehead, bugle, and hawkweed to the formula and you have a satisfying remedy for a slight liver or gallbladder problem. Prepare as previously noted.

Remedy for feverish colds, skin problems (boils, pimples, and abscesses), and children's inflammatory diseases (measles, chicken pox): Of equal parts of marigold, yarrow, safflower, catnip, and verbena, steep a teaspoon in a cup of hot water for 7 to 8 minutes, covered; let this be taken every 2 or 3 hours. (Abstinence from the usual "junk" — sugared and starchy foods — would help very much.)

Mignonette
Reseda odorata

Having company? Aromatize the sitting or living room before they arrive. Macerate 1 or 2 cupfuls of the *fresh* flowers in an open container of hot water and remove when the doorbell rings.

Spikes of the fragrant blooms will add a colorful finishing touch to your floral bouquets.

When you feel nervous or overtired, or sometimes can't fall asleep, try one of Grandfather's old-fashioned remedies. Inhale deeply the delicate scent of the flowers and *slowly* exhale.

The derivations of both the generic and common names of this plant indicate its healing value(s): *Reseda,* from the Latin, means "to heal," "to calm," since the herb was employed in ancient times as a remedy for tumors, bruise swellings, and the like. *Mignonette* is taken from the French, meaning "Little Darling," a favorite aromatic of European gardens, and is said to be derived from an Old Irish word for smooth/gentle, which in turn led to *mitigate* (pain). Thus, to bring relief to a painful bruise or swelling, add enough hot water to macerate 2 cupfuls of the dried, ground leaves and flowers, stir, and cover 2 to 3 minutes. Enclose in cloth and rub or apply as a poultice as warm as possible.

For a mild kidney problem, drink a "safe," tepid tea by infusing a teaspoon (to a cup of hot water) of all parts of the dried and cut plant, especially the roots, in a cup of hot water. To prepare an effective diaphoretic (sweating agent) in fevers and colds, drink it warm to hot every hour. Make a syrup of the liquid by syruping it with honey or sugar and take tablespoon doses every hour.

You may enjoy eating the cabbagelike flavored, early leaves of *R. lutea* (yellow mignonette or dyer's-weed). Include them *fresh* with salad greens and other potherbs. This species also yields a strong yellow color to dye silk, wool, cotton, and linen.

Garden Uses

Let the foot-high plants spread and cover the ground. They're said to benefit roses. Do cultivate them as a border for the latter.

Mountain Ash
Sorbus americana

You've seen plenty of these small handsome trees growing on lawns as mere ornamentals. But what a shame to neglect their nutrient-packed fruits. Several students of my herb-study classes have gathered bushels of the large heavy clusters of reddish berries for food and medicinal remedies.

True, these berries are rather acidulous and tart, but that's due to their beneficial content of citric and malic acids, and especially sorbic acid, and sorbitol (the latter is employed in the synthesis of vitamin C), which can prevent scurvy, spongy gums, loose teeth, rheumatic problems, etc.

British-born folks like to gather the ripe fruits from the trees, dry the berries, and grind them into flour used to prepare bread. The fruits are the chief ingredient of their tart and tasty jellies and preserves. Scots consider the latter item a most suitable accompaniment for fowl and venison and also use the ripe fruits to make good cider and wine.

Here's a remedy for loose bowels and diarrhea that's easily

Mountain Ash

prepared in your kitchen: Steep a level tablespoon of the late-summer twigs, dried and cut, in a cup of hot water for 15 minutes and drink a cupful every hour or two as needed.

A Cold/Cough Remedy
Mix equal parts of the ground mountain ash fruits, summer-gathered leaves of linden, New Jersey tea, American ivy, verbena, and a large pinch of sage. Stir *well* one teaspoon in a cup of hot water and cover 5 to 7 minutes. Stir and strain. Drink one such warm cupful every 1 to 2 hours for feverish colds. To make a cough remedy, add enough honey or raw sugar to make a syrup and take in tablespoon doses every hour. You'll find this preparation works gently to bring up annoying and persistent mucus and phlegm.

Note
Do not confuse this tree with flowering ash (fringe tree).

N

Nasturtium
Tropaeolum majus
Common Nasturtium, Indian Cress

> *"A sluggish man should eat Nasturtium, to arouse him from his torpidity."* Pliny (XIX, 44)

Nasturtiums are one of the best paying tenants of my flower patch. A dozen years ago, I substituted a large patch of these quick growing and profusely flowering plants and other useful herbs for a second spruce tree that stole too much of the garden's cherished sunlight.

Up to fall's initial frost, the nasturtium plant should provide you with a most acceptable replacement for watercress. Wash both the early leaves and the later flowers and add to vegetable salad for unusual sparkle. You'll see immediately the

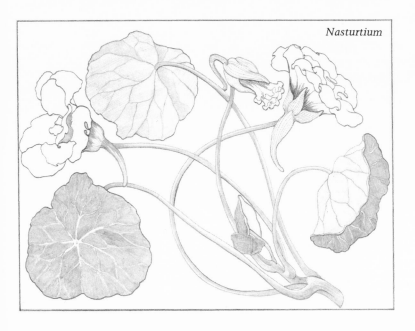

Nasturtium

connection between the herb's generic name and its pungency:
nasus and *tortus* are Latin for "convulsed" or "twisted nose."
(Watercress has a similar pungency; if you like watercress,
you'll love nasturtium.) The later leaves and the succulent
stems may be steamed and included in soups, omelets,
sandwiches, casseroles, and vegetable dishes, etc. Try putting
the leaves, their stems, and the half-ripened fruits in a jar of
dill pickles for a few days, and then add to a salad or coleslaw.
Or cover the green fruits with a hot aromatic vinegar, simmer
for 2 or 3 minutes, and let cool before bottling. Please don't
waste this health-fortifier as mere garnish.

Dry any excess or profuse growth for future meals or for a
healthful tisane.

A Nasturtium Sandwich

Gather the leaves and flowers when the latter are in full bloom,
wash and dry, and spread over a thin slice of bread. Insert a
thick slice of cheese, tuna or chicken salad, and cover with another layer of nasturtium. Cut the sandwich in quarters.

Nasturtium's edible parts go well with early leaves and
flowers of rose and violet in most dishes.

Avail yourself of nasturtium's therapeutics. Its high-
nutrient and antiscorbutic qualities equal that of watercress
and other produce, a fact well recognized by British seafarers of
the past two or three centuries. They employed the herb (as
food and tisane) to prevent and heal scurvy problems.

A Hint from European Herbalists

Express the juice or prepare a lotion by boiling the herb a few
minutes and applying it, tepid and strained, to face blotches,
blemishes, pimples, and other skin annoyances. And if you're
bothered with nasal polyps or inflammation or irritation within
the upper nostrils, use a juicer or blender to express the juice,
strain, and syringe the nostrils with it 3 or 4 times a day.

Use nasturtium as a blood purifier, in liver, rheumatic,
kidney problems, and for whatever ails you. Prepare by simmering a large handful of the cut leaves and tops in a pint of hot
water down to half the content and straining. Drink ½ cupful 4
times a day.

Do cultivate this prolific grower wherever you can — over banks and walls, in porch and hanging boxes and pots, and especially during the fall-spring period on your vacant window-sills.

By all means, interplant your squash with nasturtiums (and garlic), and you'll not lose this good food to the pesky squash vine borer. Nasturtiums will also protect your cucumber vines from aphids and cuke beetles and keep other pests away from melons.

New Jersey Tea
Ceanothus americanus

Early in 1950, S. H. Clemence of Johannesburg, South Africa, wrote me asking for some *Ceanothus* seeds and enclosed a 10-shilling note to pay for them. No, that English herbalist-chemist (a manufacturing pharmacist) wasn't plotting a Johannesburg Tea Party nor was he desirous of using the herb as a tea substitute, as had the Americans during revolutionary days. Mr. Clemence intended to raise a continuous supply of this shrubby herb in order to obtain certain extracts that he had already found most beneficial in treating spleen and liver troubles.

For everyday home use, cut, dry, and coarsely grind the overground shrub during the blooming time in June and July. A tepid tea of the leaves (a teaspoon to a cup of hot water) can become your health drink, caffeineless and very similar to regular tea in color and taste. Alone, honey-sweetened and taken tepid-cool every hour or two, the infusion becomes a strong febrifuge in feverish colds and a sedative-expectorant in whooping cough and assorted bronchial discomforts. A simmered or concentrated (boiled) tea of the parts plus the root, taken cold every 2 to 3 hours, becomes an antispasmodic and astringent in diarrhea, an effective gargle for sore throat, bleeding gums, and cankers, and a topical lotion for skin sores and irritations.

Undertaking a wool-dyeing project soon? Gather the large reddish roots, mash them with a hammer or mallet, and let them dry. A cupful of the coarsely broken parts boiled in a quart of hot water down to half the amount yields a stable cinnamon red dye.

Maybe this herbal shampoo won't smell as nice as Clairol's commercial product, but if you rub a handful of the young fruits, blossoms, and the outer set of floral leaves in a basin of warm water, you'll get a darn good lather. You might want to combine this with red clover and saponaria (which see).

O

Orchid
Orchis

There are orchids and orchids. Several of the more than fifty different kinds that could reside in your garden can serve you not only as beautiful ornamentals, but also as a highly nutritious food, as a medicine, as. . . . You're in business if you have these species: *O. mascula, O. masculata, O. biflora,* and *O. latifolia.* Southern Europeans and Middle Easterners cultivate these orchids specifically for these purposes.

Wash the thick tuberous roots of, say, the more popular *mascula* species, remove the outer skin, and dry them thoroughly by suspending them in the air. Cook, steam, or bake them, and consider them a nutrient-filled substitute for the potato. Or you may powder the dried root, season, and cook it in hot water to prepare a pleasing and acceptable farinaceous

Orchid

food that is especially suited for children and convalescents. A very small amount cooked with hot water or milk (1 to 50 parts) forms a thick jelly that contains vast amounts of a jellylike substance, sugar, starch, protein matter, and minerals of calcium and potassium.

I cook mine with Irish moss to prepare a kind of pudding. Yummy!

Need to allay the irritations of the gastrointestinal tract, the liver, and the gallbladder, as well as chronic diarrhea? Make a thin paste of the mucilaginous powder, or cook the root and strain, and gradually add tepid water or milk to thin it out. Drink this as often as required to bring relief.

In folk medicine, the tubers (and their preparations) were considered a stimulating tonic, aphrodisiac, and hormone-rich rejuvenator. Centuries ago the herbalists believed that since the roots were shaped like a testicle (*orchis* in Greek), the plant would empower men with the endless passion enjoyed by the handsome Orchis of mythology. More recently, European country folk have eaten them to revive their sexual vigor.

Note

If you reside in the more southern states, grow your orchids in your garden; northerners may have to cultivate the tubers in a greenhouse during the winter season.

P

Pansy
Viola tricolor hortensis

Get to know these lovely funny-faced border plants well as a wonderful food, and you'll be gathering the leaves and flowering tops after the flowers have fully bloomed. You too will enjoy eating them in a vegetable salad, with steamed vegetables, in omelets, and in casseroles. Use this okra substitute to thicken your soup or stew. Include it in a turkey stuffing. And top your puddings and gelatin desserts with a thin spread of the finely cut leaves and flowers. A royal touch!

Summertime usually calls for a refreshing cool drink. Boil a cupful of the cut fresh leaves in a pint of hot water for 2 to 3 minutes, add the juice of ⅓ lemon, 1 orange, or a tablespoon of cut barberries, and simmer for 10 minutes in a covered pot. Strain, allow to cool, and add the least amount of sugar or honey to sweeten. Chill before serving.

Candied Flowers
Gather the larger-sized flowers in the morning, wash them in cold water, and dry. Whip ½ to 1 teaspoon of water with an egg white, and dip both sides of the flowers in the egg mixture and then with raw sugar. Place between sheets of wax paper, and store in a closed container.

Make a tea of the leaves not only as a tea substitute but as a means of ridding the body of months-long accumulations of toxin wastes. *Stir well* a teaspoon of the ground dried leaves and flowers in a cup of hot water, cover 10 minutes, and strain. Sip slowly. This tisane lives up to its old-time reputation as a mucilaginous expectorant and demulcent in bronchial troubles and hoarse or sore throat. Add enough raw brown sugar or honey to sweeten. Take a tablespoon every 2 to 3 hours as needed.

Cough Remedy
Steep a teaspoon each of yarrow, balm, and hollyhock (leaves),
and 2 teaspoons of pansy leaves in a cup of hot water. Stir
every 5 minutes and when cool, strain. Make a syrup with
honey or sugar. Take 2 tablespoons every hour or two.

Prepare a poultice for a festering sore or wound. Soak the
dried leaves in warm water for a few minutes, enclose the mass
in cloth, and bandage onto the affected area.
And do include the flowers in your potpourris.
Indeed, cultivate pansies during all the fall-winter-spring
months in your windowsill garden. The seedpods of flowering
plants will yield enough seeds for all the purposes stated above
and for the entire coming year.

Partridgeberry
Mitchella repens
Squaw Vine, Twinberry, Squawberry, Teaberry

I remember when I needed a few berried stems of this small
trailing evergreen to complete my glass terrarium and found
them growing profusely a few yards beyond my house, in the
deep, damp woods beneath the pine trees. And I also remember
that October day when I startled the Girl Scouts by nibbling on
the scarlet fruits, and how I explained that Indian doctors used
a strong tea of the leaves of this herb and raspberry to facilitate
easy labor and childbirth and to correct most uterine disturbances (hence the synonym squaw vine).
I remember too how completely flabbergasted and almost
horror-stricken Uncle Fred became as I removed a foot-long
strand of the vine from the terrarium and began to chew 2 or 3
leaves. And when I proceeded to make my health drink with
the dried leaves of last fall's collection combined with mint and
hollyhock leaves (a heaping teaspoon simmered in 2 cups of hot
water for 15 minutes). Today Uncle Fred takes the same

Partridgeberry

drink — on the QT — for his kidneys "when they act up."

I also remember the time I explained to my students the Indian medicine man's observation that since the leaves are located on opposite sides of the stem, just as the kidneys are located in relationship to the spine, so must the plant be used as a kidney remedy, a fact well corroborated by modern medicine.

Passionflower
Passiflora incarnata
Maypop, Wild Passionflower

The fruits of this plant are fully matured when they're yellow/orange color and the size of a chicken egg. A month later you may eat these satisfying, cooling "apples." Do let the pulp of other passionflowers shrivel before you preserve this sweet delicacy into jellies or jams.

After several fruits have matured, gather the flowers and tops of this plant curiosity for medicinal purposes and dry them by suspension, somewhere away from the sun. No, dear reader, if anything, a warm tea of the flowering tops *calms* the "passions," i.e., it soothes and reduces nervousness and relieves insomnia. (Homeopathic physicians employ preparations of these parts as antispasmodics and sedatives in whooping cough, dysmenorrhea, and epilepsy.)

For everyday purposes, mix well bee balm, marjoram, valerian (which see), and passionflower, and steep a level teaspoon in a cup of hot water for 5 minutes. Strain and sip slowly. Repeat once in 1 hour if necessary. Best taken on an empty stomach.

Note

Despite all the exaggerated claims relating to this herb's theological associations, the notion that passionflower is a plant of the Holy Land is completely erroneous. It is native to the West Indies, Latin America, and our southern states from Virginia to Florida.

Passionflower

Pelargonium
P. odoratissimum
Nutmeg Geranium

When our family visited Aunt Mary, the children were forewarned several times: "Now don't sit on the big armchairs. They're for the grown-ups." And for many childhood years, I reasoned that old folks' talk was not for children's ears and that our place was in the kitchen. No wonder everyone talked about her sweet smelling chairs. Sixty years later, after an infrequent visit to the front parlor and a quick sit on one of the chairs, I can divulge the obvious secret of her aromatized chairs. On their backs were suspended several multicolored silk squares like miniature saddlebags, which, I later discovered, were filled with aromatics from grandfather's garden: lavender (which see), mints, lemon balm, and other sweet-smelling herbs.

A Must

Make pelargoniums, or scented geraniums, ingredients of sachets and potpourris as well as pillow, sofa, and chair aromaticizers. Use with lavender, thyme, marjoram, rosemary, dried and ground peels of lemon, orange, and tangerine, and other sweet-smelling herbs of your choice. (For a sachet formula, see Violets.)

When an afternoon catnap is in order, rest your head, cheek down, on such an herb pillow, and your siesta will be more rewarding.

In the Kitchen

Use the young leaves to flavor commercial white wines (alone or with woodruff, citrus peels, and food seasoners) and vinegars (alone or with basil, oregano, mints, etc.); in all salads and, before adding, marinate the leaves in tepid herb vinegar for 7 to 8 minutes; in all cooked dishes; in recipes of fruit jellies, jams, and marmalades; with other aromatics to flavor fish, meat, poultry, soups, and casseroles; as a substitute for cinnamon, coriander, or cardamom in baked apples, in fruit compotes, and old-fashioned candies.

As a Healing Remedy
Make the pleasantly scented tisane a healthful herbal substitute for the caffeinated Pekoe tea. Steep a level teaspoon of the dried leaves, alone or with small amounts of mint, carnation, lavender, balm, and other aromatics, in a covered cup of hot water for 10 minutes. Stir, strain, and sip slowly. The tea is a simple home remedy that has a dual role: It helps greatly to prevent and to correct temporary conditions of stomach cramps, dyspepsia, colic, and flatulence. In nervous disorders and occasional sleeplessness, combine equal parts of pelargonium leaves with hops, chamomile flowers, valerian, or lady's slipper root, verbena, and a little sage and gentian; and in fevers and colds combine with verbena, yarrow, ageratum, and elder flowers to raise internal heat and gradually induce perspiration, thereby reducing the fever.

Peony
Paeonia albiflora, P. officinalis

Ancient healers dedicated this plant to Paian (or Paeos), "that excellent Physition . . . who first found out and taught the knowledge of this herb into posteritie."* In early Greco-Roman times, its roots were much employed in nervous and mental disorders. And in Elizabethan times, the herb was used not only for this purpose but also for headaches and convulsions. It was also considered useful as a paregoric or anodyne teething ring in the form of a necklace of beads of peony roots, which was worn by young English children cutting their early teeth. (If you use the roots today, be sure that they're *thoroughly* dried.) The practice continued during the last two centuries. (Did not American children up to the 1930s wear beaded necklaces of Job's tears — actually seeds — for similar purposes?)

*John Gerard, *The Herball or General Historie of Plants,* 1597.

In the "It's-Nice-to-Know" Department
Contemporary medical textbooks vouch for the plant's old-time
and recent use as an antispasmodic (in epilepsy), in convulsions, nightmares, spasms, and disorders of the head and
nerves. It is also recognized as an alterative in treating skin affections, ulcers, and rheumatism. Practitioners of homeopathic
medicine recognize its efficacy in vertigo, pain in wrist, fingers,
knees, and toes (arthritic symptoms), varicose veins, and
chronic ulcers.

Take a lesson from the Chinese and Japanese people who
have cultivated peonies, *P. albiflora,* not as garden ornamentals
but as everyday vegetables. Treat the roots like parsnips, steam
or cook them, and benefit from their minerals (malates and
phosphates of potassium, sodium, and calcium), sugar and
starch.

Peppermint
Mentha piperita
Brandy Mint

There are literally dozens of ways of using this multipurpose
plant, of which a whole tome could (and should) be written: as
a guardian of the garden's vegetables, as a flavorer of and substitute for Pekoe tea and ingredient of a tisane (herb tea), as a
food seasoner, as a major ingredient of medicinal remedies . . .
and on and on.
Plant one or two roots in loose, moisture-retaining soil
near cabbage and *all* its relatives — cauliflower, broccoli, brus-
sels sprouts, turnips, etc. (or strew the ground mint leaves and
stems around them) — and you'll protect the cabbage and the
other vegetables from the pesty white cabbage butterfly.

Mint Jelly Variations
a. Simmer gently for 20 minutes a tablespoon of the
crushed dried leaves and 1 tablespoon of ground orange rind in

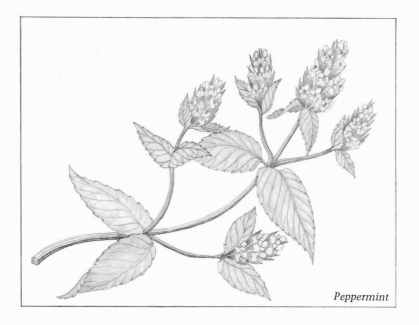

Peppermint

2 cups of prepared syrup (1½ cups of sugar dissolved in a cup of water). Strain, mix with apple, berry, or other jelly, and refrigerate.

b. In a glassful of commercial jelly, simmer ½ cup of washed, freshly gathered leaves for 6 to 7 minutes. Strain and reset. Add enough green (vegetable) coloring to yield the desired intensity.

c. Bring to a gentle boil 1 cup water, ½ cup malt or cider vinegar, 3 cups sugar, a cup of ground fresh leaves and stems, and a few drops of green coloring. Add ½ (small) bottle of liquid pectin and quickly boil ½ minute. Strain through cloth into hot, sterilized glasses. Pour melted paraffin over the jelly and seal.

d. Prepare an apply jelly base by rapidly cooking 3 pounds of sour apples, washed and sliced, for 15 minutes, and strain. Bring the yield of juice (about 5 cups) to a rapid boil, suspend a few sprigs of mint, let it boil 3 to 4 minutes, and remove the herb. Gradually add the sugar (3½ to 3¾ cups) and green coloring, stirring continually, and boil until 2 drops hang on the side of a cold spoon. Strain into hot sterilized glasses.

e. Boil 3 cups of cider, a tablespoon of lemon juice, a teaspoon of the rind, and ½ cup fresh leaves for 15 minutes. Strain, stir in the required amount of pectin, and boil. Gradually add 4 cups of sugar, and simmer a few minutes. Add 2 ounces of mint vinegar, and strain.

f. Dissolve ¾ cup of sugar in ½ cup of apple cider and 1 cup water and gently bring to a boil. Add 2 level tablespoons of the leaves, green coloring, and ½ bottle of fruit pectin. Stir constantly before and during the boiling process. Boil to full rolling for one minute, remove from fire and skim. Pour into hot sterilized jars and seal.

g. Add a peppermint flavor to various jellies by placing a few large leaves on the bottom of the jars and then pouring in the hot jellies.

Use mint jelly with cheese, poultry, eggs, meats, and with desserts.

Peppermint Vinegar Variations

A suitable flavoring vehicle for fruit salads and punch and an ingredient of mint sauce for lamb, veal, and other meats.

a. Use 1 to 2 cups of fresh mint, washed and toweled, ½ to 1 cup of sugar, and a pint of cider vinegar. Bring the vinegar to a boil and add the sugar and leaves, stirring. Simmer 6 to 8 minutes, strain, and pour while hot into hot sterilized bottles; seal.

b. Or dissolve the sugar in the vinegar, add the mint, and bring to a boil. Simmer, strain, and pour.

c. Pour the vinegar syrup onto the mint, cover the jar tightly, and let steep two weeks. Strain.

d. Gather the leaves before the flowers appear and pack them into a jar. Add enough cold or hot cider vinegar to fill the jar ¾ full and let stand covered for 2 weeks, etc.

Mixed Herb Vinegar

1 quart cider vinegar
1 cup peppermint
handful of lemon balm
handful of tarragon
½ tablespoon lemon peels, cut
½ tablespoon orange peels, cut
2 or 3 cloves
3 one-inch cinnamon sticks

Warm the vinegar and pour over all the ingredients. Seal and let stand in a warm place, stirring daily. Strain.

Pat your minted vinegar onto mosquito bites and other skin itches. Shake 2 ounces of vegetable oil and 6 ounces of mint vinegar, and apply to sunburn. For that sore bruise, prepare a liniment of equal parts of the vinegar, turpentine, and witch hazel extract. Shake well and rub in vigorously.

To prepare a gargle, stir a heaping teaspoon of dried sage in ½ cup of mint vinegar and cover 10 minutes. Strain and gargle warm every 1 or 2 hours.

For skin eruptions or hives, prepare a thin paste of enough corn starch mixed into 2 ounces of mint vinegar. Apply every hour.

Mint Sauce Variations

a. Mix together a cup each of the finely cut leaves and cider vinegar, simmer 10 minutes, and cover 30 minutes. Stir in 2 ounces of honey and serve.

b. Cover the finely ground leaves in enough boiling water to soak, and let stand until cold. Mix with marmalade and use with fruits or desserts.

c. Make a syrup of a tablespoon of sugar or honey in ½ cup of cider vinegar and pour over ⅛ cup of minced leaves. Cover and set the container where it will keep warm. Let it infuse an hour, and strain.

Use a mint sauce with a fruit cup or dessert, venison, poultry, and lamb.

Mint Butter

Simmer a handful of the leaves and parsley in very little water. Drain, rub through a Foley food mill or similar apparatus, and work into 4 ounces of unsalted butter.

Mint Pasty

Mix well equal amounts of chopped peppermint leaves and currants and a thin spread of honey. Use as a filling in pies and turnovers instead of apples. Such pasties are quite popular in northern England.

Mint Punch Variations

a. Mash a quart of fresh leaves, washed and dried, and cover with boiling water for 10 minutes. Add 2 cups of cold grape juice and ¼ cup each of lemon and orange juice. Sweeten with sugar to taste and add a quart of ginger ale. Either add ice cubes to the bowl or fill each glass ⅓ full of ice and fill with the punch.

b. Crush a large handful of mint over which pour ½ to ⅔ cup of orange juice (freshly expressed is better), and add ¾ cup of sugar. Add a pint of cider, 1½ quarts of ginger ale, and enough coloring to yield a nice green, and stir.

c. Prepare an infusion of peppermint and spearmint by covering the leaves with hot water. Let stand overnight and strain. Mix with 8 ounces of pineapple juice, a pint of apple juice, and the juice of a lemon. Add enough sugar or honey to sweeten. To serve, dilute with an equal amount of water.

Mint Julep Variations

a. Chop a handful of the leaves (and flowers) into small pieces and stir them with a cup of hot water. Cover for five minutes, strain, and add the liquid to a syrup composed of a pint of water and 4 ounces of sugar. To this add 4 ounces each of orange juice and strawberry juice and the juice of 2 lemons (plus their rinds, which remove when ready to serve). Pour ½ pint of claret wine, stirring, and add chopped ice. Garnish the bowl with mint sprigs, whole or cut strawberries, and/or thin slices of orange.

b. Stir well 1 to 2 ounces of wine, 3 to 4 fresh leaves (crushed or cut), and 1 teaspoon of honey. Add ice to suit.

c. Warm 3 ounces of wine and add 8 or 10 large crushed leaves. Cover until cool, stir, and add a little crushed ice.

d. Northern Julep: Prepare a tea of dried peppermint and sage (a teaspoon of combined herbs to a cup of hot water) and add a tablespoonful of wine to each cupful and enough honey to sweeten.

Fruit Cup

Put small scoops of melon, cantaloupe, and watermelon, and diced fresh pineapple into a bowl and pour over ½ cup of sugar or 2 tablespoons of honey. To each quart of fruit, add ½ tablespoon each of peppermint and apple mint, finely chopped. Refrigerate 2 hours before serving.

Peppermint Cookies

1 cup butter	1 teaspoon vanilla
1 cup raw sugar	2½ cups unbleached whole wheat
2 eggs	flour
	2 teaspoons finely ground mint

Cream butter and sugar. Mix all ingredients. Drop by the teaspoonful on a cookie sheet. Flatten with bottom of floured glass. Bake 10 to 15 minutes at 350 degrees.

Mint Cake

Prepare your pastry the usual way and flavor with honey and chopped fresh leaves. Roll out thin and bake.

Peppermint Candy

Boil a cup of cut leaves in a quart of boiling water for 15 minutes and strain through cloth. In this dissolve 2 cups of sugar and simmer until a soft ball forms when dropped into water. Remove from heat and beat in ½ tablespoon of cream of tartar until creamy. Add desired coloring and drop spoonful amounts onto pan or waxed paper.

Herbal Lozenge

Use also equal amounts of mint, violet, linden, hibiscus (or hollyhock root), marigold, or carnation. Proceed as noted under Peppermint Candy up to using the cream of tartar. Simmer the strained solution until it begins to thicken. Pour onto a buttered pan and cut in squares.

A Minted Bath

Mix together equal parts of peppermint, pine, balm, thyme, and rosemary, and a few peels of orange and lemon. Enclose in cloth or several layers of nylon stockings and tie it directly under the water spout; or let it soak in the bath. Rub your skin with this "bag" and feel how refreshingly stimulating it is.

Aromatizing Your Foods

Mint imparts a delightful flavor to such steamed vegetables as parsnips, peas, beans, carrots, and potatoes. Add a thin sprinkle of the fresh finely cut mint over the food a few minutes before serving. Also helps to prevent possible gas after eating.

◆ Dust a little of the powdered herb over soups, stews, and purées.

◆ Sprinkle a little of the fresh or dried leaves over baked or broiled fish, and within sandwiches of cheese, tuna, and chicken. Include in hamburger and casseroles.

◆ When preparing pancakes or waffles, use 1½ to 2 teaspoons of crushed leaves to flavor 5½ to 6 cups of batter (milk, eggs, flour, molasses, butter, and raisins).

◆ When baking apples, fill the cored fruits with a little honey and dried mint.

Mint Tea

Prepare a tisane or health beverage by steeping the herb alone or with equal or smaller amounts of lemon and tangerine peels, rose, pelargonium, New Jersey tea and/or lavender. Stir a teaspoon of mint (or combination) in a cup of hot water and cover for 5 to 7 minutes. Stir, strain, and, if necessary, sweeten with honey. Sip the drink slowly. You'll love its marvelous fragrant taste and find it an excellent substitute for Pekoe tea and far more relaxing and soothing to the nervous system.

Internal Remedies

Use peppermint (and other available mints) in all internal remedies. It is a reliable antispasmodic and carminative in simple disorders of the gastrointestinal tract, the liver, and gallbladder. For this purpose, mix equal parts of mint, catnip, balm, linden, and turtlehead and drink an infusion (a teaspoon in a cup of hot water) 3 or 4 times a day.

Remedy for Nervous Headache

Mix equal parts of peppermint, catnip, valerian, lady's slipper, and balm, and steep a teaspoon in a cup of hot water. Drink warm every 2 to 3 hours as needed.

Antacid

Mix equal parts of peppermint, chamomile, linden, bee balm, carnation, and a sprinkle of dillseeds. Prepare as above. Drink a warm cupful as often as needed. For children, steep half the quantity of herbs. For infants, replace carnation with chamomile and catnip (obtainable at health-food stores).

Cold-breaker Remedy

Drink hourly warm teas of peppermint, elder flowers, verbena, sage, New Jersey tea, and a touch of crocus stigmas (saffron) or marigolds. (Most of my herbal remedies intended for internal use have contained peppermint, spearmint, curly, or other mints of my garden.)

Note

Peppermint is elsewhere referred to as "mint."

Periwinkle
Vinca minor
Myrtle, Lesser Periwinkle
V. major
Greater Periwinkle

There's more to these fast-spreading evergreens than just being goody-good ground covers for sun or shade areas. They beg to present you with an outstanding ecological benefit. You may control and forevermore prevent the erosion of those steep slopes bordering your garden and lawn by strategically placing root divisions and rooting stem cuttings every 10 to 12 inches apart.

Note that the similarity between this very binding/healing effect upon the soil and its counterpart in human therapeutics has been long established. Periwinkle and *Vinca* are from the Latin words *pervinca* or *vincio,* to bind or twine, which describes the plant's astringent/remedial properties. Indeed, use a decoction as a warm gargle for inflamed tonsils, cankers, mouth and throat ulcers, to irrigate bleeding sores and annoying crust formation within the (nasal) septa, to stop diarrhea, and as an enema for bleeding hemorrhoids. (Boil a large open handful in

Periwinkle

1½ pints of hot water down to a pint in a covered pot.) Apply a more concentrated decoction (i.e., it is boiled another 10 minutes) to assorted irritations and diseases of the skin, eczema with bald spots, and sores that refuse to heal.

In recent years, periwinkle has been hailed as a source of an anticancer drug. Bravo! But you may help prevent health disasters by drinking a "blood-purifying" tea, the ingredients of which are equal parts of the entire periwinkle plant, dried nasturtium, eupatorium, and American ivy. Simmer a tablespoon in 1½ pints of hot water for 10 minutes and let cool. Stir and strain ½ cupful 3 or 4 times a day. Sweeten with very little honey.

Do enjoy a tepid drink of the summer leaves (with mint and balm) as a refreshing tea substitute (a teaspoon in a cup of hot water).

White Pine
Pinus strobus

Here are several suggestions that put to good use your pine tree, its leaves and innumerable cones. Anticipate the Christmas holiday and get set to prepare herbal wreaths and evergreen decoctions, with which to deck home and church. Your Yuletide arrangements will require sprays of cedar, mountain laurel, white pine, and other evergreens. Create your individual motif by adding variously sized pine (and other evergreen) cones and colorful gourds. Wire the cones to the positioned sprigs of evergreens and tie the completed wreaths with red ribbon. Do fasten one to your front door for holiday cheer. Why not a wreath as a table centerpiece, or over the fireplace? A post-season hint: The handsome cone wreaths are almost indestructible and can be used for many years. Store them in a plastic bag containing one or two mothballs.

Try making miniature Christmas trees of the large-sized cones. Add that extra touch by glittering, painting, or spraying them with clear plastic, gold or other preferred color, and with

White Pine

an occasional stripe of a harmonizing shade. Both arrangements of cones and wreaths are suitable for church and home, office and work areas.

You may wish to string or wire a series of the painted cones over the fireplace, large mirrors, doorframes, and mantelpiece. How about stringing the unpainted ones on the *inside* of the fireplace, away from the flames, so that the warmth will diffuse that pleasant pine aroma? And now and then, throw one or two cones onto the burning logs.

An herb pillow? A simple matter of drying the summer evergreen leaves and stuffing them, minus the lower thick stems, into a zippered, 15-inch colorful pillowcase. Mm! Love that slightly balsamic but acceptable fragrance that lasts for many years. And it's a beautiful conversational piece.

You don't have to observe the Indians' spring ceremony of fasting for a week and drinking copiously only of teas of pine leaves (needles). (Taking such infusions several times a day has, until recently, been a common practice of northern Russians.) This, their "ounce of prevention," should be ours too to help rid the body of winter-accumulated toxins, prevent colds, and possible scurvy-associated illnesses. The needles' high content of vitamins A and C was corroborated by my friend, Dr. Harold Feldman in the 1930s. Interesting that the vitamin C content of pine leaves is 5 times that of lemons.

To prepare a healthful and refreshing tisane, gather the needles from spring to early summer, dry them thoroughly, and grind them to a ½-inch size. Mix equal parts of these leaves and your favorite aromatics (balm, mint, New Jersey tea) or food seasoners (thyme, basil, marjoram). Stir a teaspoon of the mixture well in a cup of hot water, cover for 15 minutes, stir, strain, and sip slowly.

Extra Food

Gather the seeds (the nuts) in August, just as the cones are opening. You'll find they're sweet and do not taste of pitch. Cook them in soups and stews, with fish and meat, and include them in meat loafs, fish cakes, casseroles, and stuffings.

Remember that old-time drugstore favorite, Syrup of White Pine Compound? A chief ingredient was the crude inner bark of

the pine tree. You can obtain some of it from the larger limbs
you've trimmed from the tree. Strike them hard with a mallet
or wide-headed hammer, remove the inner bark from the outer
layer, and let dry for several days. Cut them into small pieces.
For decades my herbal pharmacy sold a soothing syrup for
common bronchial disorders. Make your own by boiling in a
covered pot a quart of hot water containing a cupful of equal
parts of sassafras, hollyhock, American ivy, wild cherry, thyme,
and especially the bark of white pine. When the liquid has been
reduced to half, strain, let cool, and add enough sugar or honey
to make a syrup.

A Kidney Remedy
Mix equal parts of the bark, eupatorium, hydrangea, and saxifrage. Simmer a tablespoon in a pint of hot water for 15 minutes, strain, and drink tepid-cool 3 to 4 times a day.

Don't neglect the viscid resin that oozes from blisterlike
cavities of the tree's trunk, branches, and cones. It's a chief ingredient of various external remedies that help to quickly heal
scaly eruptions, itching sores, and cracked skin. Cover the resinous pitch, coarsely cut woody parts and/or ground-up sticky
cones with enough rubbing alcohol, or slightly simmer in
warm vegetable oil for 15 to 20 minutes and keep in a warm
place. Let stand for 10 days or so, strain, and label. (The alcoholic solution compares well with the healing efficacy of
Tincture of Benzoin Compound.) Prepare an ointment by simmering the gooey pieces of the trunk, cones, and the pitch in
melted, saltless lard for 15 to 20 minutes (2 heaping tablespoons to a cup — 8 ounces — of fat). Strain into a clean jar
and allow to cool. Good for man and beast.

Platycodon
P. grandiflorum
Balloonflower

Borrow a culinary trick from the East and use the 5- to 7-inch shoots of the platycodon plant as a flavorful, though slightly bitter, foodstuff. Cook them as a pot vegetable, in bean or pea soups, or stew them with fish or fowl.

Consider too the dried leaves and stems as a stomachic tonic in minor intestinal upsets and catarrhal complaints. The remedy: Mix a teaspoonful of the dried ground leaves and flowers with ½ teaspoonful each of lavender, dictamnus, and pelargonium. Steep the herbs in a cup of hot water for 15 minutes. Drink the tea tepid every 2 or 3 hours. The roots, dried and cut, become a sedative/astringent in diarrhealike situations. Simmer a level tablespoon in a pint of hot water for 15 to 20 minutes, remove from heat, and when it is cool, drink ½ cupful every hour until comfortable. The tablespoon of herbs may contain cranesbill, privet, and water lily.

Primrose
Primula

Don't confine primroses to the obvious places; let them thrive in more unobtrusive ways: under trees, in partial shade, on damp banks, in the crevices of rock gardens. This way, you're not wasting valuable growing space, and you may be inclined to put the "little firstlings of spring" to greater advantage.

For that infrequent headache and neuralgic pain, know that a "wine" of the flowers offers exceptionally marked sedative properties, especially as a sleep-inducing remedy. Gather the early flowers just as they're opening and let them ferment a few days in a thin syrup of sugar and water. Or cover the aromatic flowers with wine for 10 days and strain. A tablespoon is a "dose."

Primrose

Prepare a calming, refreshing tea by stirring well a heaping teaspoon of the dried, ground leaves and flowers in a cup of hot water. Cover 10 to 15 minutes before straining. This pleasant tasting drink may be taken between meals and an hour before bedtime.

The fresh early leaves of the *Officinalis* and *Vulgaris* species may be taken with salad vegetables, cooked (steamed) as a potherb, or incorporated chopped in soups and casseroles. I'm sure you'll enjoy the anise-scented and piquant taste of the leaves. Before adding the leaves to a salad, immerse them in an aromatic vinegar or in wine a few minutes to sharpen the taste.

Add the fresh flowers to all sorts of pottage, thick soups, purées, and stews of meat and fish, to cooked/steamed vegetables, buckwheat groats, rice, etc. Season with a touch of saffron, safflower, or marigolds. M-m-m good!

Privet
Ligustrum vulgare
Common Privet, Prim

Like barberry and arborvitae, this shrub pays good returns for the space it occupies. The glossy blackish berries that follow the small, sweet-scented flowers are relished by bullfinches, homing pigeons, and other caged birds.

Ask a Spanish or Portuguese grandmother what *alheña* is, and invariably she will recall that "back home" the ladies prepared a rose red dye by boiling the leaves in hot water. They also boiled radish roots and the leaves and root of privet to impart a strong orange yellow dye to cloth and a reddish tint to the hair. (For a straw shade of hair coloring, see Quince.) Pick the leaves in fall for your golden yellow dye.

For healing purposes, gather and dry the *late* summer leaves. Use a strong infusion (2 teaspoons steeped or 1 teaspoon simmered in a cupful of hot water) as an estimable gargle for sore throat, cankers, and for spongy or bleeding gums. Value its astringency in chronic intestinal looseness (e.g., diarrhea) and ulcerations of the stomach. Steep a teaspoon of the dried leaves and twigs in a cup of hot water and, when cool, drink the strained liquid every 2 to 3 hours. (See Cranesbill.)

Use the cold, simmered liquid as is to heal infected cuts and skin wounds, with an equal amount of warm-tepid water. To use as a douche in leukorrhea and vaginal discharge, dilute the strained infusion with three parts of water.

Remember
These latter healing applications are essentially verified by recent discoveries of plant researchers. The antibiotic principles of three privet varieties, they found, inhibited the activity of pus-forming bacteria.

Pyrethrum

Chrysanthemum coccineum, C. roseum
Painted Lady, Painted Daisy, Persian Pellitory, Insect
Flowers

The flowers of this plant, states the *U.S. Dispensatory*,
"are highly toxic to many lower forms of life [worms]
. . . [and] to the cold-blooded vertebrates, such as fishes and
frogs." The powdered flowers have long enjoyed a reputation as
one of the best contact poisons for insects, thus the common
synonym insect flowers.

You will find that it's not necessary for the insects to
eat the powder to be killed. Spray or dust lightly to control
moths, fleas, flies, silverfish, mosquitoes, ants, roaches, and
other home invaders. The powder can be applied similarly to
dogs and cats as a flea expellant. (Use the formula given in Arborvitae.) Suspend sprigs of pyrethrum from the ceiling and
screens of your porches and your attic. That'll shoo off those
annoying insect pests.

Fear not, pyrethrum is harmless to humans. Several
pyrethrum-containing commercial parasiticides have been mar-
keted in lotions and tinctures (alcoholic solutions) and as
fumigators. I recall that my long tressed sisters, suspected (often unjustly) of head lice, would first wash their hair, dry it
thoroughly, and then, under Mother's watchful eye, vigorously
rub a little of Grandpa's nit (insect) powder into the scalp, and
wrap the head with a towel. Then Aunt Esther or Aunt Mary
would "inspect," comb, and rebraid the tresses. (See also Feverfew.)

Q

Quince
Cydonia oblonga, Cydonia vulgaris

> *"There is no fruit grown in the land that is of so many excellent uses as this, serving as well to make many dishes for the table."* —John Parkinson, *Botanicum Theatricum*, 1640.

A "twice-told tale." One summer's day 20 years ago, neighbor John called me over to his yard. "You want these quince fruits? They're yours," he said. "I don't want them. They're too bitter." I explained that here in the temperate zone (especially in New England), these fruits are more firm than soft, more acidulous than sweet, and admittedly almost unappetizing. To no avail. "Take them," he commanded, and I did.

True, I did not then eat the fruit like an apple, although they are closely related. But with the few fruits (pomes) presented to me, I was able to prepare a dozen or more items. Days passed and with my various kitchen-made quince products laid out on my neighbor's lawn table, I proceeded to explain the virtues and applications of each.

For some years following that demonstration, John and his sisters zealously treasured each fruit.

Earn your K.C. (Kitchen Chemist) degree by preparing quince in these ways:

  ◆ Compote with other fruits—apples, pears, and peaches—using honey to sweeten.

  ◆ Bake. Brush off the fuzz, peel, and core 3 or 4 quinces. Fill the cavity of each with honey or sugar, place in a shallow dish, add water to cover the bottom, and bake-baste them alone or with apples and other fruits in a slow oven until soft. (Or use a syrup of honey and water to pour over the fruits.)

  ◆ Boil until soft, sweeten, and serve warm with cream. A healthful dessert or snack.

  ◆ Prepare a spiced syrup (water, raw sugar, cinnamon, and clove), and simmer gently. Cut the fruit in quarters, core (but save the seeds for future use), and cook in the syrup until soft.

◆ A quiddany or confection much enjoyed by royalty and the wealthy of the 1600s and 1700s is a thick fruit jelly, thicker than a syrup but not as thick as marmalade. Boil the cut and cored fruit in a thick syrup of raw brown sugar or honey until very soft, and rub through a sieve. Cook the mixture again in the syrup until stiff, and, when cool, add ground rose petals (the original recipe calls for rose water), and powdered cinnamon or aniseeds.

Marmalade (a 300-year-old recipe)
Cook whole quinces until soft, cut and pound them, and press through sieve. To each pound add ¾ pound of sugar, aromatics to taste (cinnamon, cloves and/or ginger), and stir well. Remember: Marmalade is from the Portuguese *marmelada*, a conserve of quince.

Jelly
Cut into thin slices (do not core), put into a double boiler, add a cup of water to each 5 pounds of fruit, and cook until soft. Ex-

Quince

press through sieve and proceed as with other jellies. Allow ¾ pound of sugar to each pint of juice. Tart or sweet apples may be used with the quince to yield a more pleasing flavor.

Dressing
Simmer the above jelly until it is of the consistency of cream. Add honey or sugar to sweeten and use over fruit salad.

Sauce
Include with apples and other fruits. Prepare a quart of a thin jell of quince, add a cup of cranberries, and 2 cut apples, and cook 10 to 15 minutes or until a thickened consistency results. Express or strain and, when cool, add the required amount of liquid pectin (as advised by the label). Goes well with fish, fowl, and meats.

Wine Dip
Immerse the blossoms, plus marjoram and carnation flowers, in wine for a few minutes; chop, and include in a salad.

Quince-Mint Syrup
To a previously prepared syrup of raw sugar or honey, add 4 cut and cored, not fully ripe quinces, a pint size of mint leaves, and a glassful of red roses. Stir intimately and let this steep for 24 hours. Then boil gently until it is "half wasted," strain, and add more sugar or honey to thicken (from an eighteenth-century recipe).

The seeds are equally as important as the fruits. Remove them from the cored parts and set aside to dry. They abound in a thick mucilage, which is easily extracted by boiling water, and form a most pleasant confection and remedy. Their grateful inclusion in a variety of drinks and externally used items has been a constant practice for thousands of years.

Prepare a thick jell by boiling a heaping tablespoon of the seeds in a pint of hot water for 45 minutes and, when cool, straining it into a glass jar. To remedy upset stomach, vomiting, and diarrhea, sip small amounts every few minutes. Taken as is or extra-syruped, the jell will quickly relieve coughing spells and certainly helps to remove annoying phlegm from the bronchia and stomach. In case of laryngitis or throat hoarseness, mix 1 or 2 drops of Tincture Benzoin Compound into a teaspoonful of the honey-jell base and sip slowly.

Eye Lotion
Dilute a thick *plain* jell with 3 parts of tepid distilled water. Ease minor irritations with 1 or 2 drops in each corner of the eye.

Or you may digest ½ to 1 teaspoon of the seeds plus a 2-finger pinch of barberry, marigold, or rose petals in 8 ounces of distilled water over a warm stove for 5 minutes, stir frequently, and strain.

A Facial
Begin with a thick jell and gradually incorporate small amounts of vegetable oil, 1 or 2 teaspoons to a cup of the quince, stirring well. Massage the area in the morning and allow to dry into the skin. Wipe off with a moist cloth. Good for wrinkles (this sixteenth-century "wrinkle" pomatum is quite serviceable today), rough skin, crow's feet, etc.

Or: Use equal parts of the jell and freshly expressed juice of aloe.

Hand Lotion
Use either the *plain* jell or prepare a fresh one by boiling a teaspoon of dried seeds and one of Irish moss in 1½ cups of hot water for 7 to 8 minutes. When cool, strain, and add 2 teaspoons of glycerin and enough toilet water to scent. Shake well before using.

Or:

1. Macerate 1 teaspoon seeds in 8 ounces of cold water overnight, and strain.

2. Dissolve ⅔ teaspoon of borax and ½ teaspoon of boric acid in 4 ounces of water.

3. Add 2 ounces each of rubbing (grain) alcohol and glycerin.

4. For extra preservative-protection, dissolve 8 grains of benzoic acid (obtainable at your pharmacy) in the alcohol.

5. Mix together all solutions and label.

Hair Lotion
Add enough toilet water to a quince mucilage to thin to a suitable consistency.

Flaxen Hair Coloring
In a quart of thin jell, simmer a large handful of the ground dried (thicker) twigs of privet for 15 to 20 minutes, let cool, strain, and rinse the hair. To deepen the shade, add more of the twigs and a tablespoon of chamomile flowers, and boil the mixture.

Pomander
Substitute the usual apple or orange with the fall-maturing greenish yellow quince or use its shell. Stick the entire fruit with cloves or fill the shell with peels of orange, lemon, tangerine, roses, orris root, violets, and other aromatics. Tie and suspend.

R

Rose
Rosa

When I was in my early teens, I helped my grandfather transplant scores of wild roses to our home grounds. We "tamed"
them and they grew tall, offered varicolored and fragrant flowers and at least a dozen purposeful uses, and served as a protective boundary around our garden. (You can do the same.) Thus
rose-gardening was a part of my apprenticeship as Grandpa's
disciple of landlording the tenants for their rent in return for
their space occupied on our home grounds.

   The rosebush affords a plentiful supply of ingredients for at
least a dozen homemade items. First and foremost, the preparation of sachets and potpourris requires the indispensable presence of unopened rosebuds with their long remembered, welcome scent. Gather them on a warm sunny day between 10 and
12 A.M.

Sachet
Use the buds as the base and include equal portions of dried
citrus rinds, the blossoms of lily of the valley, heliotrope, and
lilac, spices such as clove, cinnamon, and allspice, aromatic
herbs (the flowers of sweet clover, milkweed, and chamomile),
and, of course, the everyday food aromatizers — marjoram,
thyme, lavender, etc.

   Combine all ingredients and stuff into suitably sized bags.
Mix in a sprinkle of ground sweet flag or florentine orris root
(see Iris) as the required fixative.

   Keep a sachet in the corners of the living-room chairs and
sofa, in dresser drawers, and in linen and clothes closets. You'll
be delighted with the fragrance as well as its use as a possible
moth deterrent.

   Rose enthusiasts use the buds and petals to make herb pillows.

Herb Pillow
An excellent conversation piece. Stuff a zippered toss pillow

with dried pine leaves, lilac flower *buds,* and a liberal sprinkling of bee balm, pelargonium, dictamnus, and especially lavender. Needs refilling every 3 or 4 years. A superb Christmas
or birthday gift.

Potpourri

Use the same ingredients listed above under Sachet and

a. Add a few drops of perfume or preferred aromatic oil
(eucalyptus, bergamot, or rosemary, obtainable at your pharmacy) and stir well with a ladle. To a 9-ounce glassful of the
mixture add 2 teaspoons of either benzoin gum, roots of orris,
or sweet flag, or a combination.

b. To prepare a moist potpourri, alternate layers of the
above floral-foliage ingredients with coarse salt in widemouthed
glass jars, and keep weighted down with a heavy object. Stir
every third day for 3 weeks. Add fixative and other ingredients
and seal the jars to complete the seasoning.

A Rose Bath

Stuff the buds, citrus peels, safflower, dictamnus, pine, mint,

Rose

and food seasoners (marjoram, thyme, etc.) into the toe of a doubled nylon stocking. Tie it to the water spout and let warm-hot water run through slowly, or tie it to the shower spray, submerge it in the warm tub water, or use it as a sponge or washcloth, and instead of soap.

Room Aromatizer

Add a pleasant, almost tantalizing aroma to a room by stirring into a small pan of hot water a small handful of rose petals and a level teaspoon of the pleasantly scented sachet ingredients. Place the container next to or *on* a warm radiator or stove. When preparing any rose product, always use nonaluminum and noncopper cookware, spoons and stirring rods, strainers and containers.

A Vase of Roses

In a tall clear-glass vase or suitable glass bottle, prepare alternate layers of multicolored blossoms — white, yellow, pink, red — and repeat. This is an attractive and colorful ornament and a good gift idea.

A Health Tea (Tisane)

In a cup of hot water, stir well a teaspoon of the dried, ground rose leaves and 2 or 3 crushed fruits, small pieces of citrus (dried) rind, and a 3-finger pinch of peppermint, carnation, dictamnus, and bee balm. Cover 10 minutes. Stir, strain, and sip the cupful 3 or 4 times a day.

The rose's malic and tartaric acids serve to rid the gravel from the urinary tract. In my former pharmacy days, my European customers found the tea of great value in "dissolving out" gallstones; they took a cupful every 2 hours on an empty stomach. Try adding the leaves and petals to Pekoe tea.

A freshly made tea of the petals is a pleasant, cooling drink during warm summer days. Use it instead of water when making cakes, bread, pies, puddings, sauces, or gelatins. John Ronald, a member of my herb-study class, prepared throat-soothing lozenges by first simmering various herbs in rose water and making the result into a candy according to traditional recipes.

Rose's leaves, flowers, and fruits are therapeutically a mild astringent, diuretic, an aid to the liver, gallbladder, and bloodstream, a tonic/stimulant, and refrigerant (in the summer-time). The aforementioned tea, when boiled down to half its content, is a corrective for loose bowels.

Eat the Petals
They make a nice nibble, may be included in a fruit or vegetable salad, soup, or purée. To prepare them for a sandwich, butter them and refrigerate. Then add enough fresh petals to cover.

A Between-Meal Snack
Dab a fresh petal into honey or maple syrup. Ground or powdered, they can be incorporated into all cooked foods — soup, baked or broiled meats, casseroles, baked goods, jellies and preserves, sauces, fruit molds, and even omelets. Try the finely ground, dried petals in cottage or cream cheese, sour cream, or other spreads. Use alone or with parsley and/or dill.

To candy your rose petals, see Violet.

Pickled Rosebuds
Place 25 or 30 buds in a quart jar, over which pour a syrup composed of 2 cups cider or your favorite vinegar and ½ cup of sugar or ⅔ cup of honey. Seal with paraffin wax and store for 3 or 4 weeks in some secluded dark and warm place.

The buds go well, chopped, in salads, coleslaw, and as a sandwich ingredient.

Rose Water
Slowly simmer a glassful of dried petals in 2 pints of hot water in a covered pot for 45 minutes.* Express the liquid, in which mix another glassful of the petals, and either prepare as before or distill the liquid in an old-fashioned long-spouted metal teapot or similar apparatus. The distillate is obtained by placing or suspending a shallow dish or curved metal 2 inches away from the teapot's mouth. Preserve the aromatic water in the refrigerator.

---

*My grandfather usually used 3 consecutive days of sun heat to prepare his aromatic waters.

Other Uses of Rose Water
Drink diluted as a health tea (see above).
◆ Use it to flavor baked goods.
◆ Bring to a boil a tablespoon of chamomile flowers mixed in a pint of rose water, simmer 2 or 3 minutes, and let cool. Strain. Use as a skin lotion, facial, and hair rinse. (Dilute and drink for stomach cramps.)
◆ Mix equal parts of the rose water and glycerin and use for chapped or rough skin. Or incorporate as much rose water as possible into lanolin or cold cream.
◆ For a diarrhea remedy, simmer a teaspoon each of loose-strife, cranesbill, spirea, or privet in a pint of warm water for 10 minutes. Strain and take ½ cupful.
◆ To make a vaginal douche, dilute the above solution with 2 or 3 parts of tepid water or dilute the rose water with an equal amount of water.
◆ For a rose syrup, thicken the original rose water with enough sugar or honey and use as a topping for fruits and compotes, as flavoring and sweetener for Pekoe and herbal teas, as a base for jellies and jams, and as a simple cough syrup.

Rose Hip Syrup
Chop 2 cups of ripe fruits and boil in a little water until tender. Express the juice, set aside, reboil the pulp in the same amount of water, then simmer for 7 to 8 minutes and again express. Mix the 2 juices and let strain overnight through muslin (or jelly bag). Stir in the required amount of sugar or honey, boil until thick, and pour into sterilized jars.

Rose Hip Soup (Swedish)
Cover 2 cups of the ground hips with water and boil for 10 minutes. Strain and again boil the fruit remains in an equal amount of water. Gradually add a thin, watery paste of flour (potato, rye, soya, whole wheat, barley). Serve warm or cold, with or without cut cold vegetables.
A costless source of high-potency vitamins. Take advantage of rose's high vitamins C and P content by eating one or two freshly picked fruits (hips) every day, as long as they last. These two vitamin health fortifiers help prevent overall infection,

scurvylike ailments; and sustain sound gums, teeth, and clear smooth skin. The fruit's bioflavonoid content, especially vitamin P, serves to keep the tiny blood vessels strong, elastic, and rupture-free, and aids the healing of wounds. (A good reason for eating more fresh fruits.) Rose hips are gathered after the petals mature; then their nutrients are at optimum. Compared to oranges, they offer 1.2 percent protein to 0.9 percent, 28 percent more calcium, 25 percent more iron, 25 times more vitamin A, and 10 to 40 times more vitamin C. (The hips are higher in vitamin C than even tomatoes and red peppers.) Vitamins E, K, and P of rose hips also surpass those of orange.

Raisin Substitute
Cut the fruits in half, remove the seed core, and dry thoroughly. Eat them alone, with apples or berries, or with sour cream or cottage cheese.

Rose Hip Purée
Boil 2 cups of the fruits in a pint of hot water for 20 to 25 minutes or until tender, and rub through a sieve.

Rose Hip Jam
Use 2 cups of the purée, 2 apples or pears, peeled, quartered, and cored, and 1¼ pounds turbinado sugar or a pound of honey.

Cover the cut fruits with just enough water, cook until tender, and add the other ingredients, heat gently, and stir until the sugar is dissolved. Boil to thicken the syrup or until a sample tested on a cold surface forms a thin skin. Be sure to thoroughly cook the mixture to evaporate the excess liquid. Pour into hot sterilized jars and immediately seal.

Hip Jam #2
Before cooking, remove both ends of the fruits and the seeds with a stainless steel knife, cover the balance with water in a glass or enamel saucepan,* and stir in 2 ounces of wine. Cover and place in a cool spot for 3 days, each day mixing with a *wooden* spoon. Rub the pulp through a sieve. Stir in 2 cups of

*Copper or aluminum utensils destroy the vitamin C content.

sugar or 1½ cups of honey, and cook quickly to prevent the loss of vitamin C. Stir until thick, but keep the lid on when not stirring.

Hip Jam #3
Use 2 cups hips, before seeding, 1½ cups sugar, 1 cup water, and 2 tablespoons lemon juice. Prepare hips and measure 1 cup. Boil sugar and water 4 minutes, add hips and lemon juice and boil, covered, 15 minutes; uncover and boil 5 minutes more. The berries should be clear and transparent and the syrup thick. When done, pour into hot sterilized glasses and seal. If hips are very ripe before frost has touched them, add more lemon juice to each cup of hips.

Vinegar of Rose
Pour a pint of cider or wine vinegar over a cupful of tightly packed petals, contained in a quart glass jar. Seal and place the jar near a warm radiator or oil burner, or in a sunny window for 2 to 3 weeks. Shake gently every 2 or 3 days. Strain and store the vinegar in a stoppered cruet. Use alone or with other aromatic vinegars, as a marinating liquid, in sauce and salad dressing, and as a jiffy remedy for scratches, cuts, and insect bites.

Wine of Rose
To a pint of red rosebuds and petals contained in a glass jar, pour a pint of red wine, a little at a time, shaking the mixture well. Stopper and let stand 4 or 5 days; shake it twice a day. Stir and strain. Use as a flavorful marinating solution.

Face Lotion
Mix 1 part of the rose water and 2 parts of witch hazel extract. Makes a nice astringent and after-shave lotion.

Eye Lotion
Steep a teaspoon of the ground dried flowers in a cup of hot water and cover until cold, stirring occasionally. Strain carefully through absorbent cotton. Place 2 to 3 drops in each eye, 4 to 5 times a day. This is a fairly reasonable facsimile of a drugstore remedy for hay fever, Estivin (a processed infusion of rose petals). Your own kitchen-prepared drops will help allevi-

ate minor irritations and inflammation of the eyes, especially in hay fever. (See also Quince.)

Hand Lotion
Mix 2 ounces each of rose water and glycerin and incorporate a tablespoon of the freshly expressed juice of aloe (which see). An agreeably scented and mildly astringent lotion.

Cologne
Mix a cup of red rosebuds, a tablespoon each of dictamnus, rosemary, and mint, ½ tablespoon each of lemon and orange rinds, and a teaspoon of cinnamon. All are dried and coarsely ground. Cover with *ethyl* alcohol and macerate a week in a closed bottle. Shake the contents every day. Strain and label. Use as an afterbath refresher, after-shave lotion, or rub-down lotion.

Extra Notes
To hasten the opening of your rosebuds, place a lump of sugar in the vase.

To that insect bite, recent scratch, or skin irritation, apply a wettened petal.

Extra fertilizer for your rosebushes is provided by spreading finely ground, dried eggshells halfway down to and around the roots.

S

Safflower
Carthamus tinctorius

If during the past 10 years you've hopped onto the health foods
bandwagon or at least become a patron of shops selling health
foods, then you too have witnessed the skyrocketing use of
safflower oil for culinary purposes and as a preventative
medicine.

Its well-deserved popularity is due to its being an "*unsaturated* oil." In recent years medical evidence has repeatedly indicated the relationship between elevated cholesterol blood
levels (hypercholesterolemia) and atherosclerosis, and to prevent the latter condition the plasma cholesterol must be reduced. Safflower oil offers the highest percentage of essential
unsaturated fatty acids* of all vegetable oils and has been
shown to appreciably lower the blood cholesterol level in
hypercholesteremic patients. Its use, therefore, is also indicated
to prevent and treat diseases of the liver, gallbladder, and the
arterial (blood) system, e.g., especially high blood pressure and
heart ailments.

Remember safflower as "American saffron" (they're in no
way related) and you'll use it similarly. Dry and substitute the
dried yellow to orange red flowers for the true saffron (see
Crocus) to color/season bouillabaisse and other fish dishes,
soups and stews, rice (plain, timbale, or pilaf), Spanish/Mexican
dishes, baked or fricasseed chicken, and bread and bakery
items.

Remember safflower as "dyer's saffron" and you may want
to use the large heads of brightly colored flowers as a dye
source, a substitute for the genuine saffron. (*Carthamus* is from
an Arabic word, *to paint.*) You may dye silk, wool, and cotton a
light orange red, which turns a deeper red when the material is
rinsed in diluted ammonia water.

---

*That is, its high content of linoleic acid.

Remedy for Liver Torpidity and Gallbladder Trouble

Mix equal parts of yarrow, alkanet, hawkweed, hepatica, and safflower. Drink a tea every 3 hours on an empty stomach. It is prepared by steeping a teaspoon of this mixture in a cup of hot water for 15 minutes. Good also as a bitterish tisane (health tea) and to sweeten a sour stomach.

Use the flowers and leaves in children's eruptive diseases. A warm infusion (alone or with yarrow, catnip, or mint) becomes a gentle diaphoretic in measles, chicken pox, and the common cold. First mix the other herbs, sprinkle lightly with the flowers and leaves of safflower, and of this combination, stir well a teaspoon in a cup of hot water for 8 to 10 minutes. Strain and give a warm cupful every 2 to 3 hours until the desired results (i.e., profuse sweating) are produced.

As in Grandfather's day, warm safflower teas may be considered a most efficacious anodyne and diaphoretic in muscular rheumatism. The "patient" will observe their efficacy more acutely if (s)he abstains from all fatty meats, boiled and fried foods, sugars, and starches.

Tisane

Stir well a heaping teaspoon of the long, slender, colorful flowers in a cup of boiling water for 10 minutes. If necessary, add honey to sweeten.

Tisane Variations

Use equal parts of mint or balm.
- Add a tablespoon of orange or cranberry juice.
- Stir the tea with a cinnamon stick.
- Add rose petals or hollyhock flowers to the strained tea, cover another 3 to 4 minutes, and eat the petals amid sips.

Saponaria
S. officinalis
Bouncing Bet, Soapwort

One August day in 1976, a phone call came from Sidney T., "Can you have ready a pound of saponaria, whole and dried?" I stepped into my front lawn where, amid my Jerusalem artichokes, grew (and still grow) masses of these erect, course perennials. A few minutes later, an armful of the plants was drying over my (cellar) oil burner. When Sidney came to collect them, he said that a lady, recently arrived from Poland, would shampoo her hair only with saponarias.

The plant's name is derived from the Latin *sapo,* soap, and relates to its (the leaves and roots) forming a lather when agitated with warm water. This sudsing property is due to a substance called *saporubin,* which constitutes 34 percent of the dried, reddish brown root. I've used the soapy solution as a scouring liquid and a decoction as an external remedy for skin itch. For this purpose, boil 2 cups of the ground roots in a quart of hot water for 20 to 30 minutes down to half the liquid content. Strain and apply tepid. My "wash compound" consists of red clover blossoms, water lily roots (which see), and the saponaria parts.

A Challenge
Substitute this liquid for your detergent base to wash your woolens. Cleans without shrinkage. And take Dr. William Meyrick's word — the decoction "easily washes out greasy spots of clothes."

Stuff an herb pillow with the *dried* pink and slightly aromatic flowers. Use them also in sachets. (See Violet.)

Saxifrage
Saxifraga
Breakstone, Rockfoil

Saxifrage (literally, a rock-breaking herb) is taken from the Latin *saxum,* stone, and *frangere,* to break; therefore, its common synonyms breakstone and rockfoil. As the astute early herbalists observed, this plant, with its tendency to grow in stony crevices and in coarse, gravelly places, is a useful gravel-removing agent in kidney and bladder disorders.

Include the young leaves, chopped, in all kinds of salads and soups. The characteristic tang of saxifrage's tender succulent leaves is due to its high concentration of minerals. Appalachian folks eat the leaves of a wild growing species called "lettuce" as a salad green.

Use alone or with nasturtium and pansy leaves. It's a good preventive medicine — helps to ward off possible kidney problems, reduces the possible inflammation of the urinary passages, and in general invigorates the whole body.

For mild rheumatic or kidney pains, cut off a rosette of leaves at their base, dry and cut them finely. You may use them either as a tisane or as a remedy. Mix equal parts of the leaves and violets, bugle, nasturtium, pansy, New Jersey tea, eupatorium, amaranth, or mint. (Use any four of the latter eight.)

The leaves may be mixed in a juicer or blender with mint, carrot, celery, nasturtium (or watercress), houseleek, and pansy. Both processed mixtures should be diluted with 1 or 2 parts of water before taken. Sip slowly a teaspoon at a time until you've ingested an ounce. Do this 3 or 4 times a day. Refrigerate the balance.

Use the blended foods to prepare a variety of spreads and salad dressings by mixing them (or finely cut leaves) well with cream or cottage cheese and plain yogurt. Add the blends to a warming soup, stew, or casserole. (And I've had a thick leaf blend as an ingredient of bread.) Steep a heaping teaspoon of the mixture in a covered cup of boiling water for 10 minutes. Strain and drink between meals and an hour before retiring.

To prepare a kidney-stimulating remedy, simmer 2 tea-spoons in 2 cups of hot water for 15 to 20 minutes. When cool, strain, and drink every 3 to 4 hours. (See also Pine.)

Scabious
Scabiosa
Pincushion

In the past century, scabious leaves have been used as an ingre-dient for purifying blood and curing skin problems. Still used by European herbalists, scabious can be enjoyed by you. Mix equal parts of coneflower, nasturtium, black birch, and dande-lion (leaves), and simmer a tablespoon of the mixture in a pint of hot water for 15 to 20 minutes. Let cool, strain, and add only enough honey to sweeten. Take 2 tablespoons 3 to 4 times a day.

Remedy for Itch and Scaly Skin
Boil a tablespoon of scabious (the entire dried plant) with a ta-blespoon each of oxeye daisy, black birch, American ivy, marigold, saponaria, and periwinkle in a quart of hot water. When the liquid is reduced to half its original quantity, strain, sprinkle a teaspoon of quince seeds over the liquid, and cool. Strain. Rub a little liquid into the dry spots and repeat as often as needed. Store the balance in a cool place or in the re-frigerator.

Observe the diagnosis of the ancients; they said that the hair-covered stem of the scabious indicated its use in allaying a tickling or irritation of the throat. Thus did scabious come to be used as a warming diaphoretic/expectorant in bronchial and pleuritic discomforts. For that purpose, you may simmer a tea-spoon each of flax, hollyhock, hound's-tongue, and scabious (leaves) in a pint of boiling water for 15 minutes. Strain when cool and add enough sugar or honey to ready the syrup. Take a tablespoon every 1 to 2 hours as necessary. Add a teaspoon each of eupatorium, nasturtium, and saxifrage to the infusion, but omit the sweetening, and the remedy helps to cleanse the urinary system of harmful deposits.

(Do note, however, that the above preparations will produce far greater results if one eliminates from the diet many salted or fatty, man-made foods, and especially those that have been improperly prepared, i.e., fried, overcooked.)

Always associate the origin of its common or generic name and its healing properties with the Latin *scabere,* to scrape or scratch. Thus, the centuries-old repute of scabious as an external remedy for itchy scalp, mange, and scaly skin disorders. The given name from *scabiosus* (Latin) rough (hairs), may describe the gray hairs that cover certain species.

Snapdragon
Antirrhinum majus
Toad's-Mouth

Emil Lindgren had been gardening glads (gladiolus) and snaps, as he called them, for all the 30 years of our druggist-customer relationship. I often exchanged my young basil, sage, and other culinary herbs for his gorgeously flowering productions and, on one such occasion, I mentioned that Grandpa made a fly-killing liquid by cooking the upper half of the plant in milk. A few years later my research into that very subject and subsequent experimentation led me to tell friend Emil that the proper procedure was to freshly express the plant's juice (I used a juicer) and add it to milk in a saucer. Place near a window or on the sill and in the sunlight, and it was sure to attract and destroy pesty insects.

However, Emil countered with one of his own did-you-knows: In old Sweden, he said, folks made a tepid poultice for slow healing ulcers and tumors. For all skin irritations, 3 or 4 whole snapdragon plants were cut and boiled in 1½ pints of hot water for 20 to 30 minutes and allowed to cool. A cloth was saturated with the liquid and applied to the affected area.

Tut tut! Just because you experience a decided but fleeting bitter taste when you bite a fresh leaf or flower does not mean

Snapdragon

it is a poison. Consider it a stimulating bitter (tonic) for stomach and intestinal problems and for liver and gallbladder disorders. Dry a few flowers and leaves, cut them into tiny segments, and steep a level teaspoon in a cup of hot water. Mix a tablespoon of the strained liquid in ½ cup of water and take 3 to 4 times a day. Sweeten with honey if necessary.

Spiderwort
Tradescantia

Take a clue from the late Dr. George Washington Carver of Tuskegee Institute. He enjoyed the "rich flavored" parts of this plant in salad and soup. Start with the latter food (or stew or casserole), progress to steamed vegetables, then to salad. I have preseasoned my spiderwort leaves and upper stems in an herb vinegar or, if possible, in a marinating liquid for a few minutes. (Use also as noted under Saxifrage.)

Break the stem or the leaf at the stem and squeeze out the juice. The juice is a good styptic that will stem the flow of blood in deep cuts and bruises and hemorrhages of the upper nasal passages. To remedy a passing irritability of the bladder, cystitis, and other difficulties of the urinary passages, express the juice via a juicer and dilute with an equal amount of water. Slowly drink this amount every 3 to 4 hours. (Or use a blender. Dilute the extractives with water and strain with pressure. Cook the leftover with other foods such as rice or buckwheat or incorporated in a soup.)

A Simple Cough Syrup

Extract 1 or 2 ounces of the jellylike juice, dilute with half as much water, and add enough honey to thicken. Or finely chop 2 or 3 spiderworts, cover with a little more than enough hot water, and simmer for 15 minutes. Strain when cool, add the honey, and stir well. Take a tablespoon every 2 hours as required.

Spiderwort

Spirea
Spiraea salicifolia
Queen of the Meadow, Bridewort

Most spireas are mild yet well-known diuretics, kidney stimu-
lants, and antirheumatics. For this purpose, pluck and dry the
leaves and flowers when the latter are in full bloom. Grind
them coarsely and mix well with equal parts of eupatorium,
nasturtium, magnolia, and saxifrage, and add a sprinkle of gentian. Boil a heaping tablespoon in 1½ to 2 pints of hot water
down to half that quantity, and dilute ½ cup of the strained
liquid with an equal amount of water. Drink *tepid* 3 or 4 times
a day.

A similarly prepared *decoction* of only entire dried spireas
is a valuable antiseptic/astringent for slowly healing sores and
bleeding cuts. Boil 2 or 3 tablespoons in a quart of hot water
down to half. In diarrhea and like summer complaints, take a
tablespoon of the cold decoction every hour.

(Mrs. T. J. Healey, dye expert of my herb-study class, reported: "Seafaring men always took the roots along on their
voyages as a [preventative of and] remedy for dysentery; the
leaves as a substitute for Chinese tea.")

Tisane
A tea of the flowers and leaves makes a health-fortifying substitute for the imported tea and a pleasant between-meal drink.
(Or steep a teaspoon of equal parts of spirea, balm, and linden.)
Sweeten with honey and take this health drink as a thirst
quencher. But should you come down with a cold or cough,
drink the tea hot every 2 hours. After an illness, take it tepid-
cool and you'll be surprised how it'll perk you up and especially
tone the stomach and intestines.

Spruce
Picea canadensis
White or Skunk Spruce

During my herbal-pharmacy years, I sold a lot of spruce gum,
which came boxed in small amounts. But in my youth all the

boys knew the winter locations of spruce trees, and always had
several chunks of (free) hard gum as standard equipment in
their pockets. (Who could afford 5¢ for a package of gum?)

Today I teach my students to make an effective healing
salve from spruce, using the same method Grandfather learned
from his Indian friends. Here's how: Scrape off or dig out about
a tablespoon of the oozing pitch, and stir into 4 ounces of warm
melted (unsalted) lard, chicken fat, or suet. Simmer, let blend
for 5 to 6 minutes, and strain into an ointment jar. Doesn't
take much to heal recent sores and wounds.

Remember your druggist's advice when you purchase that
dark liquid Tincture Benzoin Compound as an inhalant for a
deep-seated cough? "Put a teaspoon in hot water, 'tent' yourself
with a large bath towel, and breathe in the vapors." Make your
own tincture inhalant by placing a layer of the fresh or dried
gum in a widemouthed bottle, covering with rubbing (grain) alcohol, and shaking vigorously. A few days later the gum will be
completely dissolved. Stir the mixture well and filter.

Or, you may just stir a little of the dried pitch in a pan of
hot water, tent yourself, and inhale the vapors. For even better

Spruce

results, drink a warm large cupful of marigold-balm tea before and after the inhalation.

During the years 1930 to 1950, I sold many bottles of Dr. Gray's Syrup of Spruce Gum and, since then, have taught others to prepare their own spruce cough syrup. First, ready a pint of syrup of brown sugar or honey to which are added 2 teaspoons of the spruce pitch and a heaping teaspoon each of hollyhock (root), thyme, and linden. Simmer very gently for 20 minutes, strain, and, if necessary, add more sugar or honey. The dose: 1 or 2 teaspoons every 2 to 3 hours.

You too will find it an exciting experience to prepare these two skin-healing remedies in your kitchen-laboratory. Remove the oozing gum from the tree's trunk and allow it to dry for 3 or 4 days. Place a tablespoon in a bottle (add, if available, a tablespoon each of fall-collected bugle, marigold, and hound's-tongue) and cover with 4 ounces of rubbing (ethyl) alcohol. Cover tightly. Shake once daily for 2 or 3 weeks and strain. Label date of preparation and ingredients.

Good for fresh cuts, sores, and irritations, and so are the following remedies that also heal recent burns. Slowly simmer a heaping teaspoon of the gum and 2 teaspoons each of the sumac and Saint-John's-wort flowers in 4 ounces of either vegetable oil, melted, *unsalted* lard, or suet for 30 minutes, stirring occasionally. Strain into a suitable container. Label.

Want good free starting material for your fireplace? After you've trimmed the branches of your spruce, gather them (and those of your neighbors), chop them in foot-lengths, and dry them. Mmm, what an aroma!

Star-of-Bethlehem
Ornithogalum umbellatum

Every time you eat these apple-sized and sweet-flavored bulbs, remember that two or three thousand years ago they represented a nourishing staple for the peoples of biblical and Mediterranean lands, from the Middle East to Rome and

Greece, and that they're still in use today. You may steam or bake them with other foods, or include them chopped in soups, stews, and casseroles. Have fun roasting them in an open fireplace.

Baking Bread
First dry and powder the bulbs. Sift carefully and use 1 to 2½ cupfuls of combined whole wheat, rye, soya, or other flour. Add a tablespoon of molasses and honey, 2 teaspoons of powdered kelp, 2 eggs beaten, the required yeast, and a cup of sour milk. After the kneaded mixture has risen, to full height, sprinkle enough sesame and sunflower seeds to cover the loaf and bake 1 hour at 350 degrees.

Turkey Stuffing
Use enough of the chopped fresh or dried bulbs to represent the bread crumbs or chestnuts. Preseason them by soaking them in wine for about an hour. They're especially good in a sage stuffing for that Thanksgiving or Christmas turkey.

Prepare a Savory Nut Loaf
Mix well 1 cup of the dried, coarsely ground bulbs, 2 or 3 cups of chopped vegetables, including parsley, 1 cup of thin nut butter dressing, ½ teaspoon each of marjoram, savory, and marigold (or safflower). Place the mixture in an oiled pan and bake in a moderately hot oven for 45 minutes.

The *steamed* nutritious bulbs of the plant are a must for anyone suffering with inflammation of the stomach and intestines, ulcers, etc. They're easily digested. Be sure to save and drink every drop of leftover juice.

Don't let that empty space beneath trees or shrubs go to waste. In early fall, plant the bulbs 2 inches deep in good garden soil, and water well until growth commences. Or plant them 6 to a pot of soil. They flower in late spring. Remember to lift, dry, and store them like their cousins, the hyacinths and lilies.

T

Flowering Tobacco
Nicotiana

Good to grow, good to know, and good to use around the garden.

Protect your flowering plants from harmful, annoying crawling insects. Prepare your own spraying solution by stirring a large handful of the dried *mature* leaves, preferably ground or cut, in about 3 or 4 quarts of hot water for 15 minutes. Allow to cool completely and strain.

Don't discard your tobacco wastes (leaves and stems) in the fall. Their high content of nitrogen, phosphorus, and potassium makes them an excellent organic fertilizer. Mix with other organic (plant or vegetable) materials and either add them to your compost pile or spread thinly as a mulch.

Grandfather called tobacco leaves his "summer remedy."

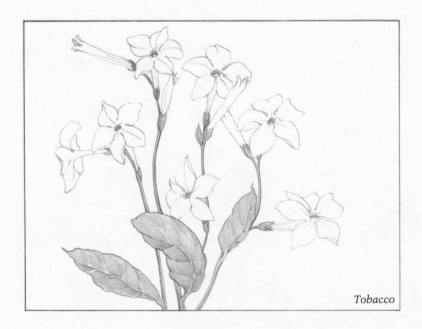

Tobacco

His small bucket of rainwater guarded the back-door entry and served primarily as a handy thirst quencher. But he'd use some of it to make a jiffy remedy to relieve the hurt of a recently acquired painful bruise, slight swelling, or beesting. You can do the same. Boil a handful of the dried, cut leaves (with a little of hound's-tongue and marigold flowers) in a pint or so of water for 10 minutes, remove and apply as a warm poultice every 2 or 3 hours.

Tulip
Tulipa gesneriana

A Case of Mistaken Identity
When tulips were first introduced into Holland nearly 400 years ago, a merchant of Antwerp received from Turkey a bale of cloth in which was enclosed what he believed was a present of onions. Some were cooked and eaten by the merchant, who so enjoyed the delectably flavored food, he had his gardener plant some in his vegetable patch. When the bulbs later materialized in the form of flowers of resplendent red, he realized that the tulips had traveled in disguise and had found a new home with him in Antwerp, Belgium.

A Century Later
The case of the sailor whose breakfast cost a huge fortune. He had delivered certain goods to a Dutch businessman and was rewarded with the morrow's breakfast, which included smoked herring. The sailor reached for what he thought was an onion and ate it with the fish. What he had eaten was a bulb of tulip Semper Augustus worth $1,500. That meal cost him more than a series of royal, entertainment-filled banquets would have!

Today's tulips are plentiful and quite inexpensive so that you should try eating these wholesome, nutritious bulbs. Add them sliced or diced to a soup, stew, or casserole, and especially to prepared chicken or turkey (baked, curried, fricasseed, broiled, etc.). In a 2-egg omelet, you may include ½ tablespoon each of the chopped bulb and green pepper, and just before you turn the omelet, add a teaspoon of cut (fresh) nasturtium, early maple (leaf), and chives, plus a pinch of basil or oregano.

Tomato Sauce

2 tulips, minced fine	2 level teaspoons dried
1 clove garlic, speared	minced basil and
with a toothpick	1 teaspoon thyme
3 tablespoons vegetable oil	1 teaspoon sea salt (option-
4 cups tomatoes, stewed	al)

Cook the tulips and uncut garlic in the oil a few minutes and remove the garlic. Add the other ingredients and gently stew for 50 to 60 minutes or until thick.

Substitute tulip for onion in mushroom and other sauces, in meat and fish loaves, rice and buckwheat (groats) dishes, and wherever onions are to be cooked or sautéed.

Tulip Tree
Liriodendron tulipifera
Whitewood, Yellow Poplar, Tulip Poplar

Now really! You should not be so constantly troubled with stomach and liver discomforts, what with this beautiful, stately tree adorning your lawn. At flowering time, gather and dry any number of the erect and scented flowers, leaves, and end twigs. Steep a heaping teaspoon in a cup of hot water and slowly sip the infusion 2 or 3 times a day. If you drink this tea, you will lose much of the phlegmatic matter from the mucous lining of the gastrointestinal tract, you'll feel fit as ever, and your appetite and energy will be restored.

The following recipe is intended as a mild stimulating tonic to the aforementioned area and the liver: Mix 1 ta-

Tulip Tree

blespoon of the dried and ground leaves and flowers with an equal amount of any four of the following: hepatica, alkanet balm, nasturtium, maple, and hollyhock; add ½ tablespoon of gentian and a light sprinkle of crocus (stigmas) or 2 teaspoons of marigolds. Gently simmer a level tablespoon in a pint of hot water for 15 minutes, let cool, and strain. Sip slowly ½ cupful every 3 to 4 hours. Add a *little* honey to sweeten.

Turtlehead
Chelone glabra
Salt-Rheum, Balmony

What's in a name, you ask? Back in Grandfather's day, this plant was better known as salt-rheum * and, without benefit of

---

*Rheum, from the Greek word flux or flow, and denoting a condition resulting from toxic, watery (mucous) wastes.

academic research and scientific analysis, was used by him and his herbalist-cronies to cure rheumatism and "wandering [rheumatic] pains," i.e., bursitis and gout.

So, when Uncle Henry or Aunt Rose agree to take your "herb cure" for their sporadic rheumatiz, gather the upper half of the plant, especially the leaves and flowers, when the latter are in full bloom. Dry and grind the parts to infusion size. Mix with equal parts of any four: false bittersweet, hollyhock, nasturtium, valerian, hydrangea, and eupatorium.

Boil 2 tablespoons of the mixture in 1½ pints of hot water for 15 to 20 minutes. Taste the strained tepid-warm tea before they drink it, and better add a little honey to conceal the strange flavor. Have them take 2 tablespoons in a little water 3 or 4 times a day.

And for other relatives, young and old, who'd like a near sure-cure for their persistent stomach distress, indigestion, and liver/gallbladder complaints, let them try a warm-tepid infusion of the dried leaves and flowers, with balm, linden, and dictamnus. Mix equal parts and steep a teaspoon in a cup of hot water. (Tactfully advise them, however, that the equally persistent cause of their digestive distress may well be dietary, that they should refrain — as much as possible — from fried and overboiled foods, high-caloried starch and sugar no-nos, i.e., assorted pastries, spaghetti, bread, mashed potatoes, pudding, etc., and the usual commercially spiced and salted eatables.)

Remind teen-agers Sue and Bill that salt-rheum is a well-named synonym for eczema, an inflammatory condition of the skin that is said to often result from long-standing dietary mishaps. Better to heed the aforementioned dietary warning so that the herbal (or other) remedy may efficaciously clear up their eczemalike skin problem.

Wait, one more synonym, balmony. The dried plant may be simmered in (unsalted) lard or suet to prepare a healing ointment, or balm, which helps to soothe skin inflammations and relieve the pain of ulcers and bruises. Use ½ cup of the ground dried leaves and stems to 1 of an ointment base.

V

Valerian
Valeriana officinalis
Garden Heliotrope, Cherry Pie

Poor Dorothy Stearns! Gardener extraordinary and poet laureate
of our herb club. How she loved and cared for every resident of
her vegetable and flower patches. Her garden heliotropes flowered so bountifully and were a gracious contribution to the
once bare and ever-wettening stone wall. And then came time
to further their propagation. One noontime, halfway through
transplanting the cut roots, she sat down to lunch and then a
short nap, when suddenly ———! She took a quick look into
the garden and even quicker steps to the rear door, which she
opened and closed very quietly. Stealthily she tiptoed into the
garden to behold in complete bewilderment an army of cats and
their cousins and their aunts and other felines — of all colors,
shapes, breeds, and sizes — all congregated atop and especially
below the wall.

There, crowding about the once 2-foot-high mound of
roots, so Dorothy described the scene, another vast assembly of
cats was feasting on the maladorous roots. Some were nibbling,
others merely content to sniff them and rub their faces against
their target; still others were rolling about in the grass or
pouncing about — drunk? — following their root-orgy, and then
jumping back to the top of the wall.

Cats aren't the only ones who have a fondness for valerian
roots — so do mice, rats, and probably other rodents. No wonder
European rat-exterminators have used valerian-scented bait in
their traps. Maud Grieve, England's foremost contemporary
herb expert, has stated that "the famous Pied Piper of Hame-
lin" owed his irresistible power over rats to the fact that he
secreted valerian roots about his person.

To describe the foul-smelling roots as "strongly aromatic"
is to point to what William Shakespeare wrote of a close rela-
tive, the *V. phu*, commonly known as "Cretan spikenard": [It
has] "the rankest compound of villainous smell that ever of-

fended nostrils" (*Merry Wives of Windsor*). Phu, the specific name, has become the equivalent of today's phew (phoo), a slang expression of disgust, a contempt that characterizes the pungent and strongly offensive odor of most valerians. It's interesting that the Romans burned the plant for incense and used the ground roots to scent linens and clothes and to ward off moths and vermin; also that the blossoms of this plant actually have a pleasing fragrance, which deteriorates on downward to overt disagreeableness.

The everyday folks of medieval England considered valerian's flavor a delightful addition (possibly the leaves and flowers) to broths and pottages. John Gerard remarks in his *Herball* that housewives so venerated the herb that a (cooked) food was worthless without it. The roots and their oil, now deemed exceedingly disagreeable, found acceptance during the sixteenth century as a perfume and are still used in Eastern lands as a constituent of perfumes and the oil of soap perfumery. (To the Digger Indians the thickened roots of *V. edulis* became, after prolonged cooking, a staple food.)

Hold Valerian's Healing Virtues in High Esteem

Use the dried roots, finely cut, to correct most of your stomach-intestinal disorders and kidney and bladder complaints and, more especially, to calm the system in most nervous disorders. During World War I the English successfully used an extract of valerian in treating shell shock and overwrought nerves and, centuries before, American Indians used the decoction of the roots to calm emotional excitement and functional nervousness. Here's how:

a. Steep ½ teaspoon each of valerian roots, washed and dried, mint (or catnip), marjoram, and verbena.

b. Mix equal parts of valerian, lavender, lady's slipper, skullcap, motherwort, and chamomile. (Obtain the latter three ingredients where herbs are sold.)

Procedure for both remedies: stir a teaspoon of this mixture 35 times in a cup of hot water. Cover for 15 minutes. Stir, strain, and drink tepid 4 times a day. Works better if you've eaten most sparingly during the day.

Equally good for "change of life," neuralgia, and simple nervous unrest.

Dr. Oliver Wendell Holmes, America's outstanding philosopher-physician, praised its worth: "valerian calmer of hysteric squirms."

During my herbal-pharmacy years, I heeded the advice of herbalist Nicholas Culpeper (c. 1640) and prepared a remedy that quite often would relieve painful irritations and spasms of coughing and shortness of breath. I followed his suggestion and boiled a tablespoon each of the roots of valerian and licorice, aniseeds, and mallow in 2 quarts of hot water down to half measure. The strained liquid was syruped with sugar and the preparation sipped in tablespoon doses every hour or two, or as needed.

For a poultice to relieve the pain of fresh bruises, a sore spot, or swelling, enclose the fresh leaves of valerian, a few sprigs of arborvitae, and wormwood in a large water lily leaf. Soak it in *hot* water for 4 or 5 minutes, and apply as warm as possible to the affected part.

Verbena
V. hortensis, V. hybrida
Garden Verbena

How true that "there's more to a flower than meets the eye." (Sorry, author unknown.) This plant's name originates from two Celtic words meaning "to drive out" and "stone," which immediately pinpoints its therapeutics. It's really no wonder-of-wonders cure-all, but if you'll drink a tea of it and other "stone" preventers, you'll note in time that the quality and tone of your gallbladder, liver, and kidney tract has improved a thousandfold.

The Remedy
Mix equal parts of alkanet, hepatica, nasturtium, saxifrage,

marigold, and hollyhock, and stir a heaping teaspoon 30 times in a cup of hot water. Cover for 15 minutes, stir/strain, and sip slowly a cupful 4 times a day.

Cold/Cough Remedy
Replace the saxifrage and hepatica with eupatorium and fever-few. Also, for that feverish cold and accompanying chills or fever, steep a heaping teaspoon in a cup of hot water. Drink the tea every hour as warm as possible. For a sore throat, gargle it warm as needed.

Cough Syrup
Simmer a tablespoon in a pint of hot water down to half and strain. When cool, add enough raw brown sugar or honey to thicken. Adults may take a tablespoon every hour or two and children under twelve half that amount.

Spring (and Everyday) Tonic
Stir well a teaspoon of a mixture of equal parts of verbena, bee balm, mint, carnation, gentian, and American ivy in a cup of hot water and cover for 15 to 20 minutes. Slowly sip a cupful every 2 to 3 hours.

Veronica
Veronica
Speedwell

Attend my herb-study classes and invariably you'll hear me explain how the Indian medicine doctor, a veritable healer and doctor, recognized and applied the therapeutic virtues of a particular but *nameless* plant to a specific disease. His diagnosis would indicate that a plant that is covered with hairs signifies its usefulness in all tickling feelings. That's true of veronica.

To remove the slimy accumulations that irritate the throat, gargle every hour with a warm-hot decoction—a tablespoon of the finely ground flowering parts and leaves boiled in a pint of hot water until only half the liquid remains. You'll also obtain

quick relief from annoying catarrh of the nasal passages by sy-
ringing that area with a tepid solution of half-and-half decoction and water.

Alfred Ruttiman, my old-time mentor, teacher, and dear
friend, knew the *officinalis* species as an all-healing perennial
and was quick to recommend *Grundheil,* as he knew it in
Europe, "for whatever ails you. . . . See how it thrives in the
rock garden, so it must be good to remove stone from the
body." And that's exactly how the clue-seeking Indian herbalist
doctors would associate the plant's stony habitat with its
ability to remove gravel formations from the kidney stones.

Kidney Remedy
Of equal parts of veronica, saxifrage, eupatorium, spirea, and
nasturtium, prepare a tepid tea (a teaspoon to a cup of hot water) and drink it 3 or 4 times a day.

Instead of rushing out to buy a cough remedy, why not
prepare your own? It'll contain no harmful drugs or chemicals,
as noted on the label. Simmer a teaspoon each of veronica,
hepatica, thyme, hollyhock, and bugle in a pint of water for 25
to 30 minutes. Cover till tepid, strain, and drink ½ cupful
every 2 or 3 hours or as needed. In case of a persistently irritating cough or a catarrhal problem of the stomach or bronchia,
add enough sugar or honey to syrup the infusion and sip a tablespoon every 2 or 3 hours.

A jiffy external remedy for a teen-ager's pimples: Boil a
large handful of the dried plant in a pint of hot water down to a
third, let cool, and strain. Sop on with a cloth 4 or 5 times a
day. It's also very beneficial for slow healing sores.

Viburnum
V. opulus and V. americanum
Cranberry Bush, High Cranberry

This moisture-loving shrub is a shining example of an all-
purpose and utilitarian resident of the garden or lawn — a triple

source of nourishment, healing, and especially, beauty and charm. It is best known as cranberry bush or high cranberry, though not related to the bog cranberry. Ever since the fifteenth century, when Geoffrey Chaucer in one of his *Canterbury Tales* suggested that the "gaitre-beries — shal be for your hele" [lit. heal, thus "to heal you" or "for your health"] and to "picke hem right as they grow and ete hem in," the bright red clusters of the bitterish tart fruits that ripen in August have been substituted for real cranberries to prepare a sharp-flavored jam or jelly, a piquant sauce, or other fruit-containing products.

Jelly

Put the fruits in a kettle and barely cover with water. Simmer 30 minutes and strain. Use as is or combine with an equal amount of elderberry, barberry, wild cherry, mulberry, or blackberry juice similarly prepared. To 2 cups of juice add 2 cups of sugar and boil until a true jelly forms. Store in sterilized sealed jars. The jelly has a beautiful color and delicious flavor. Use with meat, fowl, or fish, or prepare according to an elderberry or other fruit jelly recipe.

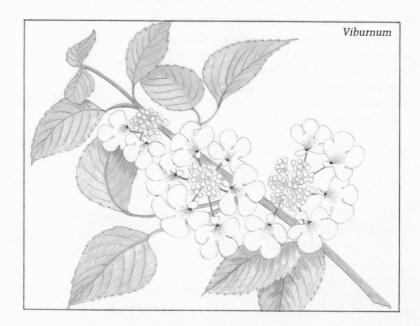

Viburnum

Spice the Fruits
Prepare a syrup by boiling 3 cups of sugar, 1½ cups of herb-seasoned cider vinegar, and ½ to ⅔ cup of water for 10 to 15 minutes. Immerse a cheesecloth bag containing 3 or 4 short sticks of cinnamon, a teaspoon of whole cloves, 3 or 4 thin slices of lemon (or a teaspoon of dried lemon peels), and boil another 10 to 15 minutes. Let the syrup cool 10 minutes, add 1½ quarts of the washed fruits, and slowly heat to 180° for 10 minutes.

Serve cold as a condiment or garnish with meat and salads.

Spicy Sauce
Cover a cup of the viburnum fruits with water and cook until they're soft. Express the pulp and make sure to exclude the large flat stones. Mix with 1 cup of blueberries, ½ cup of sugar, ½ teaspoon of powdered cinnamon, and ¼ teaspoon each of nutmeg and cloves. Bring to a boiling point and continue to boil 5 minutes, stirring occasionally. Serve warm with puddings, ice cream, and gelatin desserts.

Be sure to include the fruits in preserve and sauce recipes of barberry, crabapple, lemon, and orange.

Wine Them
Fill a quart jar half full with the fruits (plus barberries, crab apples, and others), add ⅔ cup of sugar, and add ½ cup of brandy and enough wine to fill the jar 2 inches from the top. Seal and store in a dark spot for a month or two before using.

The wined fruits go well with chicken or fish.

Pickle the Fruits
Combine 2 tablespoons of molasses, 2 cups of cider, malt, or herb-flavored vinegar, ¼ cup of honey, ⅛ teaspoon each of cinnamon and nutmeg, and stir in a cup each of blueberries (or huckleberries) and viburnums. Let the mixture stand 4 or 5 days or until enough drawn juice has covered the fruits. Pour into sterilized jars and seal.

Make Raisins
Dry any excess on trays either in the sun or in a low-heated oven. Turn them over occasionally and heat the other side until they're fully dried. Store in a thoroughly dry jar.

To restore them to fruit size, add only enough warm water to cover.

You're hardly expected to strip your viburnum shrub of its bark just because it is medically recommended for the following purposes. But you may substitute the late summer leaves and twigs as an ingredient of diuretic, uterine sedative, and nerve tonic remedies.

Leave it to the Indian herbalist-doctor to inform the early colonists *why* the viburnum was such an effective kidney stimulant — because it grows along streams and wet locations. Mix equal parts of mint, corn silk, saxifrage, hydrangea, and viburnum, and steep a teaspoon in a cup of hot water. Drink it tepid 4 times a day.

For periodic pains and cramps, mix equal parts of valerian, lady's slipper, gentian, hawthorn, and viburnum. Prepare as above but add a pinch of cloves and cinnamon to the tisane before straining, and drink it warm. (The druggist's H.V.C., Hayden's Viburnum Compound, contains viburnum, wild yam, and prickly ash, and until very recently was prescribed by physicians for its antispasmodic and nerve-sedative virtues in female disorders. The viburnum used in H.V.C. is known as cramp bark.)

Free Tea
Mix equal parts of pelargonium, balm, peels of lemon and orange, marigold, and the viburnum leaves and twigs of midsummer. (Mint and other aromatics may also be included or substituted for these ingredients.) An excellent thirst quencher.

Sorry, bird lovers. The fruits are not relished by our feathered friends, but those of the *prunifolium* species (black haw, stagbush) are greedily gobbled up by thrushes, robins, and woodpeckers.

Sweet Violet
Viola odorata
Florists' Violet, Garden Violet

> *Marie F:* "What? Eat my beautiful violets? I wouldn't
> think of it."
> *Me:* "But their taste is so delicious."
> *Marie F:* "And destroy the looks of my garden?"
> *Me:* " —and they digest as easily as lettuce."
> *Marie F:* "Heavens, no!"

(Three years later and Mrs. Marie F. is a most zealous advocate of eating violets, nasturtiums, hollyhock, and other flower-garden residents.)

When my four sons were small, I'd conspire with my wife to include the fresh, early violet leaves in a green salad. Ah, the trick—the leaves were first dipped in an herb-seasoned wine/vinegar solution (equal parts). And to her cooking vegetables, and to a bean or pea soup, she'd include the leaves and flowers and call it the gumbo soup of the day. Another variety of violet, *V. esculentus* (wild okra), is even more jellylike and is used in the South to thicken Creole soups and stews, and cough syrup.

Violets have played a part in political intrigue for hundreds of years. When Parisians wore violets around the 1830s, they were indicating their allegiance to the Liberal party.

Before leaving Elba, Napoleon I vowed he'd return with violets, i.e., in the violet season. From then on he was secretly referred to by his adherents as "Corporal Violette," and the flower thence became the emblem of the Imperial Napoleonic party. His followers recognized each other by wearing violet-colored rings or watch ribbons. Following his escape from Elba, his friends welcomed his return with bouquets of violets.

Even after Waterloo, and the replacement of Louis XVIII on the throne, it was dangerous to sport violets in one's buttonhole; this was considered a seditious act.

> *Farewell to thee, France! but when Liberty rallies*
> *Once more in thy regions, remember me then —*
> *The violet still grows in the depth of thy valleys*
> *Though withered, thy tears will unfold it again.*
>
> Lord Byron

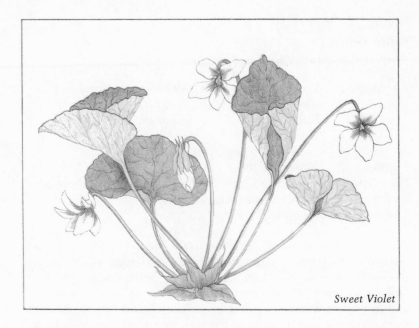

Sweet Violet

Violets are well worth the space they occupy in your garden. Their cultivation is simple, and they multiply quite profusely. They'll remunerate you with food, food seasonings, a delightful scent, and many healing remedies.

Green Beans and Violets

2 cups green beans
1 tablespoon vegetable
  oil
¼ cup sunflower seeds
1 tablespoon fresh
chives, cut

½ tablespoon chervil
¼ teaspoon each of basil,
  marjoram, and savory
1 tablespoon fresh violet
  leaves, cut

Steam the beans 10 minutes. Heat oil in a heavy skillet and add the sunflower seeds, chives, chervil, and herbs. Stir well and add the beans and violets.

This recipe suggests that you may include violet leaves with other foods to be steamed, stewed, or casseroled. Include them too in an omelet, in fricasseed chicken, and in fish and meat patties.

Violet Vinegar

Half fill a widemouthed pint jar with the *fresh* flowers and add a handful of dried and cut orange peels. Cover with warmed cider and malt vinegar for a week. Let stand in a warm place for 10 days. Shake the contents daily. Strain and label.

Use the vinegar in salad dressing, sauce, and as a marinade base.

Herbed Sour Cream (or Yogurt)

Finely chop a teaspoon each of chives, parsley, chervil, dill (leaves), sorrel, rose petals, and violet leaves. (Optional: minced greens of garlic and onions.) Incorporate in a cup of sour cream. Use as a spread or for hors d'oeuvres.

Herbed Cottage Cheese (or Ricotta)

Moisten a heaping tablespoon of the *mixed* vegetable ingredients of the previous recipe, finely cut, plus a teaspoon of marjoram and mint, with a mixture of equal parts of vegetable oil and herb vinegar. Blend a pint of cottage cheese and a small package of cream cheese and mix in the other ingredients. Refrigerate at least a day before serving.

Aromatic Wine

To a pint of white port wine, add a cupful of fresh violet flowers and ½ tablespoon each of carnation, rosemary, thyme, lavender, and marjoram, and a teaspoon of whole cloves. Let stay in a warm place, next to a warm radiator or an oil burner, for 7 to 10 days. Shake the container daily.

An excellent adjunct to a sauce, salad dressing, and marinating liquid (for poultry, meat, and fish). Use it too to apply to scratches and insect bites.

Candied Violets

Wash and dry thoroughly ½ cupful of the stemless flowers.

Beat 2 egg whites with 2 teaspoons of water until ultrawhite and foamy. In this, dip the flowers enough to coat completely. Or you may use a thin solution of acacia (gum arabic) prepared by dissolving ½ teaspoon and ½ cup of sugar in hot water. Then dust with confectioners' sugar. Let dry on sheets of waxed paper and store in an airtight container.

Sachet

Mix well 2 parts each of rosebuds, violet flowers, and pelargonium leaves, and 1 part each of lavender, heliotrope, lily of the valley, lilac, and saponaria. (Use at least two of the latter four.) Stuff the mixture into 2 by 4 cloth bags and sew together or close the open end with string.

Hang them in the pockets of clothes, in clothes closets and drawers, and in the corners of upholstered seats.

Other ingredients: rinds of lemon, orange, and tangerine, coriander, mint, marjoram, basil, cloves, cinnamon, mace, and other food seasoners. (See Iris.)

Violet Honey Syrup

Simmer a cupful of 3 parts of the flowers and 1 part of rose petals in 2 cups of water for 15 to 20 minutes. Strain, add ½ cup of sugar and 2 tablespoons of honey, and reheat to thicken. (Save the flowers and rewarm them when cooking vegetables.)

Use on French toast, waffles, pancakes, and to replace maple syrup; and on puddings, gelatins, and other desserts.

To use as a cough syrup, mix 2 teaspoons of this preparation and ½ teaspoon of lemon juice with about ⅔ tablespoon of warm water. Sip slowly 2 teaspoons every hour or two as needed.

Violet Glace

Dissolve 1½ cups of sugar in ½ cup each of water and grape juice. Stir in 2 tablespoons of washed petals and bring to a boil. Simmer 10 minutes, let cool, and strain. Add the juice of 2 lemons and 1 orange, stirring. Pour into ice-cube trays and freeze.

Violet Sandwich

Spread a thin layer of violet flowers *and* leaves over the main ingredient: chicken or tuna salad, cream or cottage cheese, Swiss or Muenster cheese. Best if the bread is thinly sliced. Try a mixture of violet, nasturtium, and rose, and season with marjoram, basil, crocus, or marigold.

Violet Tisane

Steep equal parts of the (fresh or dried) flowers and leaves, alone or with such acceptables as pelargonium, rose, balm, or carnation. Stir a teaspoon to a cup of hot water and cover for 7 to 8 minutes.

The solo tisane is the *tisane des violettes* of European health spas which, you'll find, is a most suitable substitute for Pekoe tea. It, or the combination, can be your mild emollient/ expectorant in minor bronchial affections, throat irritations, and feverish colds. Make into syrup with honey or sugar.

Skin Healers

1. Crush, wet, and apply the leaves to the affected area.
2. Slowly simmer a heaping tablespoon of the ground, dried summer leaves and *fresh* flowers in ½ cup of vegetable oil for ½ hour. Strain and use as is, and to dress sores, scratches, insect bites, and assorted skin irritations.
3. Poultice a slowly healing sore. Moisten with hot water a mixture of the crushed or cut *fresh* leaves and a sprinkling of flax (leaves or seeds). Contain the mixture in cloth and apply it to the sore spot. Hold it secure with another cloth or plastic material.

Need a litmus substitute in your kitchen laboratory? Prepare an infusion (tea) of violet flowers and strain. Add an alkali and the color turns yellow green; acid will yield a red color.

John Middleton's *Receipt Book* (1734) informs us that if lemon juice is added to a syrup of violets, "this will make them look red; if you put in [violet] Juice [to a syrup of sugar and water] and Water, it will make them look green. If you will have them all blue, put in the juice of Violets without the Lemon."

W

Wake-Robin
Trillium erectum
Purple Trillium, Birthroot, Bethroot

A reminder dated June 28, 1942, from Richard C. Potter, director of the Museum of Natural History where I served as curator of economic botany. "Ben Charles: Please include wake-robin in your next radio series."

Here's a simple home remedy for the enterprising gardener: Save those discarded leaves and roots and dry them. Cover 1 or 2 large handfuls with rubbing (grain) alcohol for 2 weeks and shake the contents once daily.

Or simmer the above parts in melted, salt-free lard (1 tablespoon to a cup) for 20 minutes, stir occasionally, and strain.

Apply either preparation to scratches, sores, insect bites, and beestings.

Wake-Robin

Water Lily

Water Lily
Nymphaea odorata
Water Cabbage, Pond Collard

The common water cabbage and pond collard are synonyms for the springtime. Gather the larger-sized, round leaves that float on leaf rafts (pads) and substitute them for lettuce in sandwiches. Their aroma is faintly reminiscent of the deliciously fragrant flowers; the taste of the fresh leaf is not unlike that of cabbage. And it's more filling and satisfying than lettuce. You can also cut these leaves and steam them with other vegetables, cook them in soups, stews, and casseroles, and use the finely diced leaves in meat or fish cakes. Do the same with the unopened fragrant flower buds. When pickling other items, include the buds. Makes a good spicy relish.

When you're making sachets, add the summer flowers of the water lily, for their perfume will enhance the preparation. Collect them just as they begin to open and dry them *thoroughly*. (See Rose and Iris for recipes.)

Having company and you'd like to decorate the dining-room table with a full-blown water lily? Pick an unopened one in the morning, keep it in the dark, and, according to Willard N. Clute (*Botanical Essays*, 1929), "it will open whenever brought to light."

Gather the roots in late summer and dry them. They can be used to dye your (white) cloth and woolen materials a fairly stable brown. The early to midsummer roots make a good soap substitute, alone or with red clover blossoms, saponaria, and yucca (which see).

You may use the therapeutically active roots in various ways. Mash the fresh fall leaves and roots and grind/cut them finely. Apply them as a poultice, hot, to swollen, inflamed glands and, cool, to sores or abscesses. Prepare a dandy healing ointment by boiling the dried, ground roots in unsalted lard for 15 minutes (a handful to a cup of lard). This is most efficacious in treating skin disorders, scalds, and sunburn.

A decoction of the parts, i.e., a heaping tablespoon of the parts and 2 tablespoons each of sage, aquilegia, or geranium, boiled in a pint of hot water for 20 minutes in a covered pot, is a good gargle for a sore or inflamed throat and for mouth cankers. Minus the sage, the decoction, in doses of 1 to 2 tablespoons every hour, checks diarrhea.

Chief Charlie Sun-Hawk told me that warm teas (infusions) of the summer roots were highly esteemed by American Indian women for bronchial disorders, that a diluted decoction of the fall collection was much used as a vaginal injection for leukorrhea, and that a strong decoction was inserted rectally for hemorrhoids.

Wormwood
Artemisia absinthium

"As bitter as Wormwood and Gall" (Jer. 9:15). Despite this well-known biblical reference, you'll discover that wormwood will "sweeten" many conditions of ill health or "bitterness." Its bitterness then is a blessing in disguise. The herb helps to rid a sour gastrointestinal tract of harmful matter, hardening mucus, and toxins. Know that *bitter* is spelled *better,* and you'll no longer let this peculiarly aromatic member of the perennial sunflower family with its silvery gray herbiage provide only ornamental contrast to the more colorful and smaller-sized residents of your flower garden. Profit from the many uses to which the extremely bitter herb can be put.

Wait for the *second consecutive* sunny day (early afternoon) in July and September (inclusive) to collect the upper half of the flowering plants. I've always dried mine by suspending them, stems up, preferably in the attic, or near and over an oil burner. Strip the stems of the leaves and yellowish green flower heads and store them in a tight container to avoid the absorption of moisture.

Kitchen Uses
You may substitute wormwood for its cousin tarragon, *A. dracunculus,* using 1/3 the quantity in any given recipe; and for sage, in a 1 to 5 proportion.

During Lent and Easter, let the *early* ground leaves replace sage and tansy in buns, cookies, and other baked goods. ("Wormwood cakes are good for the cold stomach and to help digestion" — *The Queen's Closet Opened* by W. M., cook to Queen Henrietta Maria, 1665.)

Sprinkle a 3-finger pinch of the powder over a roast; incorporate ½ teaspoon in your poultry stuffing.

More adventuresome souls have added the dried *early* leaves to omelets, soups, and stews, and have substituted it for tarragon in pickling liquids, sauces, and vinegars. Bravo!

Tisane (Herb Tea)
Mix 5 teaspoons of the leaves and flowers and 2 teaspoons each of any acceptable aromatic (carnation, pelargonium, lavender,

etc.) and a demulcent (hepatica, violet, hollyhock, hibiscus). Stir well a level teaspoon in a cup of hot water, cover for 10 minutes, and strain. If necessary, add very little honey.

Gradually increase the amount of wormwood and lessen the other ingredients. And if you're having a cup of warm water *alone*, again start with ⅕ or ¼ teaspoon to a cup and stop at ½ teaspoon. And, again, you may mask with honey the now familiar, if unusual, pungency of warm water's bitter brew.

How and where the tisane benefits: It tones the stomach and intestinal tracts and the nervous system, corrects weakened digestion and flatulence, and problems of the liver, gallbladder, and urinary system. And it's a good invigorating overall tonic for post-illness loss of appetite.

Cheese Flavor

Take a hint from New England farm folks. Perhaps you're not a cheesemaker like them, but you too can flavor your dairy products. First, infuse (stir 35 times) 2 teaspoons of the leaves and flowers in a cup of hot water and cover until cool. Strain into a quart of milk to prepare the curds. Dip slices of your store-bought cheese in the solution and drain to dry.

Incorporate a little of the tea into your cottage or cream cheese. I have flavored plain yogurt and sour cream with a faint sprinkle of the powder.

About two dozen years ago, Mrs. Waldo Welch advised me that "folks back home" (England) treated the hurtful bruises and swellings of soccer and Rugby players with a hot fomentation of the entire plant. (A heaping tablespoon of the leaves and tops to a cup of vinegar.) Use the preparation hot to ease the painful swelling of bruises and sprains, and for muscle stiffness. (See Arborvitae.) Use it warm or cold to quickly relieve recent insect bites, beestings, and similar skin irritations.

Wormwood Vinegar

Place a tablespoon of the parts in a bottle, add a pint of warm cider or malt vinegar, and shake well. Stopper and let stay on top or next to a warm radiator, stove, or oil burner for 5 or 6 hours. Remove and let cool. Strain and label.

Other ingredients: garlic, citrus rinds, pelargonium, gentian, marigold, lilac, food seasoners (e.g., marjoram, basil), and choice spices.

Combine an equal part of this vinegar and an aromatic one of your choice and use as a marinating or basting liquid for heavily fatted meats and fish.

Use the vinegar as a pickling base for cukes, Jerusalem artichokes, cauliflower, nasturtium pods, etc.

Salad Dressing
⅓ cup wormwood vinegar
1 cup vegetable oil
1 tablespoon honey
1 clove garlic, finely
    minced

⅓ teaspoon paprika
¼ teaspoon each of basil,
    oregano, marjoram

Combine the ingredients. Shake well and warm very gently for an hour. Let cool and refrigerate several hours or overnight. Remove and let it reliquefy. Shake well before using.

Sour Cream (or Yogurt) Dressing
1 cup sour cream
2 teaspoons honey
2 teaspoons wormwood vinegar
Blend the ingredients and serve.

Sugar of Wormwood
Warm a pint of rose water and stir in a teaspoon of wormwood leaves and flowers, 2 teaspoons of aniseeds, and 4 teaspoons of peppermint. Bring to a boil and simmer 15 minutes. Strain and add enough sugar to make a syrup of the liquid. Boil it, stirring, until it becomes "candy" when dropped on a cold surface.
(From a recipe of W. M., cook to Queen Henrietta Maria, 1655.)

Wormwood Wine
Stir 2 large cups of the dried flowers in a quart of white wine or Sauterne. Let digest for 3 months. Shake the vessel every third day. Let it settle for a week and decant, filter, or strain through several folds of cloth.

Use the wine as an ingredient of your marinating liquids, to make "bitter" sweet wines, as a stimulating tonic (a teaspoon in ½ cup of tepid water 3 or 4 times a day), and as an external application as noted under Wormwood Vinegar and Wormwood Poultice.

Vermouth
The name is derived from the German *Wermut,* wormwood. It is a white wine in which are steeped the above and other sweet and bitter herbs and spices (e.g., angelica, elder, orris [i.e., iris], sage). To prepare, use the ingredients and proportions as noted below in Liqueur. Decrease the amount of syrup, i.e., the sugar, honey, and fruit juice. You may "fortify" the strained wine with brandy.

You may take the product via a mixed drink, or use it as a *bon appétit* excuse, or as a bitter tonic with warm water. It's said to be a takeoff on the spiced wines prepared by Hippocrates.

Liqueur or Cordial Variations
a. Mix ½ tablespoon each of rosemary and peppermint, ½ tablespoon each of aniseed, cinnamon, yarrow, and wormwood, flowers only. Prepare other blends with sage and lemon balm, caraway, cloves, cardamom, fennel, lavender, allspice, various citrus peels, and your favorite herbs and spices.

b. Add ½ to ⅔ cup of the mixture to a pint of brandy (or white wine) and let stand for 3 weeks. Shake 2 or 3 times a week. Strain.

c. Prepare a syrup of sugar or honey and water, add an equal amount of fruit juice, and add this to the prepared liqueur.

Poultice
Cover 2 handfuls of the dried, cut plant with hot water, stir, and cover for 5 minutes. Apply as a hot fomentation to bruises, swellings, and sprains.

Flea Chaser
Suspend a few bunches of whole fresh or dried plants in the

corners of your porch, alongside the rear and front screen doors, and from the rafters or ceiling of your attic (where they'll also repel moths).

For your dog or cat, mix equal parts of wormwood, lavender cotton, arborvitae, feverfew, and/or pyrethrum. Rub the powder vigorously close to the hair roots. Be sure to spread the powder in the animal's rest box.

Rug Protector
Grind or powder a mixture of wormwood and the other components listed above in Flea Chaser. Place a little here and there under the rug after you've vacuumed it. "Where chamber is swept and wormwood is strowne,/No flea for his life dare abide to be knowne."*

Room Deodorizer
Stir a heaping tablespoon of the above ingredients into a pint of hot water. Place the container on a warm stove, radiator, or even on a sunny windowsill. Note how well the rising vapors remove the undesirable odors from the kitchen, sickroom, and smoke-filled rooms.

You may also mix the above ingredients with those for sachets and stuff them into 5-by-5 cloth bags. (Also use juniper berries, thyme, and rosemary.) Tie them securely and place on or next to a warm radiator.

In the Garden
Plant or transplant a few wormwoods near your growing carrots and turnips, or strew the ground leaves and flowers along the drills before sowing the seeds to help protect the foods from harmful flies and bettles. You may also spray a solution of the whole plant *lightly* over your garden plants once a week.

*Thomas Tusser, *July's Husbandry*, 1577.

Y

Yarrow
Achillea millefolium

After Alice learned that yarrow — she called it by its Latin name *Achillea* as she did all her garden residents — would ward off a variety of beetles, ants, and flies, she planted rows of yarrow as borders to her vegetable and flower gardens. But why near the culinary herbs, she asked, when they need no protection? Because, Alice, yarrow greatly increases production of their aromatic oils.

Not only should you consider this hardy perennial as a most suitable border plant, but also use it as a ground-covering grass substitute, especially where nothing else will grow.

Gather the early sprouting greens, the finely segmented leaves, wash and include them fresh (whole or chopped) in salads, soup, stew, and casseroles. Chop them fresh and use as a highly mineralized and pungent substitute for chives and chervil, in spreads, cottage and cream cheese, sandwiches, etc.

Consider the dried early blooming flowers as a faint replacement for cinnamon or nutmeg in baked goods, and as indicated under Wormwood.

Label all kitchen-made items. Note ingredients and date of preparation.

Salt Substitute
The summer-gathered leaves and flowers, once dried and finely powdered, can become an ingredient of an herbal powder that is a good salt substitute. Mix equal parts of powdered and sifted yarrow, dill, basil, marjoram, peppergrass, nettles, carrot leaves, and wormwood early leaves. Combine with an equal amount of powdered kelp.

Use the powder in place of harmful salt and spices, in all cooked foods, on toast, and in sandwiches.

Yarrow Vinegar
a. Add a cup of dried leaves and blooming flowers to a pint

of warm cider or garlic vinegar. Let stand 2 weeks, stirring
every third day. Strain and label.

    b. Mix a tablespoon each of yarrow, basil, lemon rinds
(dried), and oregano, 1 teaspoon of rosemary and sage, and 1
diced clove of garlic. Cover with cider or malt vinegar, shake
well, and let stand in a warm place for 2 or 3 weeks. Shake
daily. Strain.

    c. Shake well a tablespoon each of yarrow, lavender, dill
leaves and flowers, marjoram, rosemary, and marigold in a pint
of vinegar that has been warmed. Let this stand where it's
warm for 2 weeks, shaking the container every other day.
Strain.

Cordial
Mix equal parts of yarrow, rosemary, carnation, dictamnus,
lavender, marigold, lemon and tangerine peels, and ⅓ part of
nutmeg, allspice, and cinnamon. Add ½ cupful of the mixture
to 1½ pints of wine, add ½ cup of sugar, and bring to a boil.
Simmer 10 minutes and strain.

    Or: In 1½ cups of hot water dissolve ½ cup of sugar, add
the other ingredients, and boil 15 minutes. Strain into the wine
and simmer 5 minutes.

    This is also called spiced wine. It may be taken as an
aperitif or used as a base for a marinating solution with which
to preseason fatty meats, fowl, or fish. Include all yarrow parts
in your wine and beer recipes.

Hair Rinse
Mix equal parts of yarrow, sage, rosemary, lavender, and linden
(leaves). Simmer a heaping tablespoon in 1½ pints of boiling
water for 20 minutes in a covered pot. Stir, cover, and let cool.
Follow the shampoo with a rinse and *massage well.* Especially
helpful for scaly and itching scalp.

Bath Mixture
Add balm, pine, chamomile, lavender, and other favorite
aromatics to the above ingredients in Hair Rinse and place
them in 6-by-6-inch muslin bags. Close the opening and tie securely to the spout or let it steep in the warm water at least 10
minutes or until you step into the bath. Use the bag to sponge,
not rub, the body. Very soothing to the nerves.

Face Pack
Coarsely grind marigold, linden, daffodil, hepatica, and freshly gathered yarrow, cover with hot water, and slowly simmer for 10 minutes. (Or use a large handful — whose? — of only yarrow leaves and flowers.) Enclose in cotton material and apply to the skin. Let stay 10 or 15 minutes and reapply.

Face Lotion
Simmer a tablespoon of the above facial constituents in 2 cups of hot water down to half the quantity. Then add a teaspoon each of rosemary and nasturtium. Cover until tepid-cool and strain. Apply with cotton pads to greasy skin, acne areas, blemishes, etc.

Hand Lotion
Use the above face lotion as the water base for Quince Lotion (which see).

Healing Lotion
Vigorously boil 2 tablespoons of the dried-out plant in a pint of hot water, covered, for 10 minutes. Let cool and strain. Stops all external bleeding and is a good styptic, and thus its value for all cuts, scratches, and wounds. (If Achilles could heal the wounds of his soldiers during the Trojan War, so may you.) Dilute the lotion equally with warm water and use as an enema for bleeding piles and itching hemorrhoids; dilute the lotion with 2 parts of tepid water and use as a vaginal douch for leukorrhea.

Healing Ointment
To 4 ounces of melted, unsalted lard or lanolin (or equal parts of both), add a heaping teaspoon each of the gum and twigs (or barks) of black birch, hemlock, spruce, and pine; veronica and hound's-tongue; and a tablespoon of yarrow leaves and flowers. All parts are dried and finely cut.

Simmer for 20 minutes, stirring, and strain into a suitable jar. Label date of preparation and ingredients.

Yarrow Tea
Steep a level teaspoon in a covered cup of hot water for 10 to

15 minutes. Stir and strain. Drink ½ cupful 3 or 4 times a day.
What's it good for? For almost all ailments. "A thousand leaves
[hence a synonym, milfoil] equal a thousand uses": in feverish
colds and coughs, in eruptive diseases of children like measles
and chicken pox (see Crocus), in liver, gallbladder, kidney, and
especially in blood disorders. (Its high mineral content im-
proves the quality of blood.) Good for Grandma's rheumatiz
too.

Aromatic Yarrow Tea
Mix equal parts of lavender, pelargonium, ground citrus rinds,
and yarrow flowers, and steep a teaspoon in a cup of hot water
for 10 minutes. Take every 3 or 4 hours. Helps to correct temporary derangements of the stomach and intestines.

Nervine
To the above mixture, add alyssum, viburnum, and a thin
sprinkle of valerian. Prepare as above for Aromatic Yarrow Tea.

Eye Lotion
Mix well equal parts of chamomile, fennel (seed), marigold, and
yarrow, and stir well ⅓ to ½ teaspoon in a cup of hot boiled
water. Cover till cool, stir, and filter through filter paper. Use
as eyedrops or lotion every 2 or 3 hours. Refrigerate and warm
before using again.
    To prepare a compress for sore, tired eyes, mix equal parts
of the cool lotion and witch hazel extract and apply a wet cloth
to the closed eyes.

Note
Achilleine, one of yarrow's extracts, effectively stops the flow
of blood and shortens its clotting time; several years ago it was
considered an anticancer agent.
    Tests made with extracts of yarrow's leaves and flowers
proved effective against the staph germ.

Yucca
Y. filamentosa
Adam's Needle, Bear Grass, Spanish Bayonet

While visiting Southern California and New Mexico in 1969, I was privileged to observe Indian women gather yucca roots and proceed to wash their clothes and "shampoo" their hair. The Navajo use huge quantities of the roots, I found, to wash their rug materials prior to dyeing them.

To prepare a soap ($99^{44}/_{100}$ percent yucca), dig out the root, discard the outer layer, and prepare thin slices. Then agitate a few of them in tepid-warm water to yield a foam. Shampoo the hair and vigorously massage the scalp. Rinse the hair thoroughly and dry the hair, again massaging well.

I found too that our Southwest Indians still use the fresh root as a wash for various sores, ulcers, and other skin discomforts. We non-Indians should do likewise.

Just as my make-use-of-everything cohorts and I have enjoyed eating the fresh flowers of yucca, so might you too. Gather a cluster of the creamy white, bell-shaped blooms, rinse

Yucca

with water, remove the excess moisture, and include a few in a salad. You may also steam or bake them with other foods. Add them chopped to soup.

Let the flowers add something new to your macaroni dish. Chop the flowers down to size and cook the pasta and blooms with tomatoes and pimiento — a Latin American idea.

Z

Zinnia
Z. pauciflora

Let the flowers of this favorite edging plant provide you with handy dye material. America's bicentennial year increased the interest of local homemakers and gardeners in engaging this creative practice, just as the early colonists had done.

When the zinnias are in full bloom, snip off the yellow orange heads and let them dry for a day. Use 3 to 4 tumblers of them alone, coarsely ground, or combine 2 or 3 tumblers with one containing equal parts of onionskins, barberry's woody parts, safflowers and/or marigolds. Boil in 1½ quarts of hot water ½ to 1 hour or until the desired shade is reached. Use alum as a mordant. You're now ready to dye wool, cotton, linen, and rayon in varying yellows. To obtain different shades, add ammonia water, rusty nails, tin, and barks or twigs from oak trees, sumac, hemlock, and seed hulls of various native nuts.

BIBLIOGRAPHY

BAILEY, L. H. *Standard Cyclopedia of American Horticulture.* New York: Macmillan Company, 1910.

BAKER, SAMM SINCLAIR. *Miracle Gardening Encyclopedia.* New York: Grosset and Dunlop, 1961.

BARDSWELL, FRANCES H. *The Herb Garden.* London: Adam and Charles Black, 1911.

BOERICKE, WILLIAM. *Homeopathic Materia Medica.* New York: Boericke & Runyon, 1936.

BRITTON, NATHANIEL LORD, and BROWN, ADDISON. *Illustrated Flora of the Northern United States and Canada.* 2nd ed. New York, 1913.

BROWN, O. PHELPS. *The Compleat Herbalist.* Jersey City, New Jersey, 1865.

BUCHMAN, DIAN DINCIN. *The Complete Herbal Guide to Natural Health & Beauty.* Garden City, New York: Doubleday & Company, 1973.

BYRD, ALFRED GRAF. *Exotic Plant Manual.* East Rutherford, New Jersey: Roehrs Company, 1970.

CLUTE, WILLIAM NELSON. *Plant Names.* New York: E. P. Dutton & Co., 1932.

COATS, ALICE M. *Garden Shrubs and Their Histories.* New York: E. P. Dutton & Co., 1964.

COON, NELSON. *Using Plants for Healing*. Great Neck, New York: Hearthside Press, Inc., 1963.

CULPEPER, NICHOLAS. *The English Physician*. London, 1826.

DIETZ, MARJORIE J., ed. *10,000 Garden Questions Answered by 20 Experts*. Garden City, New York: Doubleday & Company, 1974.

EVELYN, JOHN. *Acetaria*. Brooklyn, New York: Brooklyn Botanic Garden, Women's Auxiliary, 1937.

FAURETTI, RUDY F., and DEWOLF, GORDON P. *Colonial Gardens*. Barre, Massachusetts: Barre Publishers, 1972.

FERNALD, MERRITT LYNDON, and KINSEY, ALFRED CHARLES. *Edible Wild Plants of Eastern North America*. Cornwall-on-Hudson, New York: Idlewild Press, 1943.

FOLEY, DANIEL J. *Garden Flowers in Color*. New York: The Macmillan Company, 1945.

GERARD, JOHN. *The Herball, or Generall Historie of Plants*. London, 1597.

GRIEVE, MAUD. *A Modern Herbal*. 2 vols. New York: Harcourt, Brace & Co., 1931.

GRIFFITH, R. EGLESFELD. *A Universal Formulary*. Philadelphia: Blanchard and Lea, 1859.

HARRIS, BEN CHARLES. *Better Health with Culinary Herbs*. Barre, Massachusetts: Barre Publishers, 1971.

_____. *The Compleat Herbal*. Barre, Massachusetts: Barre Publishers, 1972.

_____. *Eat the Weeds*. Barre, Massachusetts: Barre Publishers, 1971.

_____. *Kitchen Medicines*. Barre, Massachusetts. Barre Publishers, 1968.

HEALEY, B. J. *A Gardener's Guide to Plant Names*. New York: Charles Scribner's Sons, 1972.

HOTTES, ALFRED CARL. *The Book of Perennials*. New York: Dodd, Mead & Co., 1958.

HYLTON, WILLIAM H., ed. *The Rodale Herb Book.* Emmaus, Pennsylvania: Rodale Press, 1974.

KEELER, HARRIET L. *Our Northern Shrubs.* New York, 1903.

KOCH-ISENBURG, LUDWIG. *Garden Guide.* New York: Viking Press, 1966.

KRIEG, MARGARET B. *Green Medicine.* New York: Rand Mc-Nally & Company, 1964.

KROCHMAL, ARNOLD and CONNIE. *A Guide to Medicinal Plants of the United States.* New York: Quadrangle Books (New York Times Book Co.), 1973.

LEYEL, C. F. *Cinquefoil.* London: Faber and Faber, 1951.

———. *The Elixirs of Life.* London: Faber and Faber, 1948.

———. *The Magic of Herbs.* New York: Harcourt, Brace & Co., 1926.

LINDLEY, JOHN. *Flora Medica.* London, 1838.

LOEWENFELD, CLAIRE, and BACK, PHILIPPA. *The Complete Book of Herbs and Spices.* New York: G. P. Putnam's Sons, 1974.

LUST, JOHN B. *The Herb Book.* New York: B. Lust Publications, 1974.

MAUSERT, OTTO. *Herbs for Health.* San Francisco: published by the author, 1932.

MEYER, JOSEPH E. *The Herbalist and Herb Doctor.* Hammond, Indiana: Indiana Botanic Gardens, 1934.

MEYRICK, WILLIAM. *New Family Herbal.* London, 1740.

MILORADOVICH, MILO. *The Art of Cooking with Herbs and Spices.* Garden City, New York: Doubleday Company, 1950.

PARKINSON, JOHN. *Botanicum Theatricum.* London, 1640.

———. *Paridisi in Sole.* London, 1629.

Pharmaceutical Recipe Book, The. 3rd ed. Washington, D.C.: American Pharmaceutical Association, 1943.

PHILBRICK, HELEN, and GREGG, RICHARD B. *Companion Plants and How to Use Them.* New York: Devin-Adair Co., 1966.

Physicians Desk Reference. Oradell, N.J.: Medical Economics, Inc., 1965.

PORTA, GIOVANNI BATTISTA DELLA. *Natural Magick* (1658). New York: Basic Books, Inc., 1957.

POTTER, SAMUEL O. L. *Materia Medica, Pharmacy and Therapeutics.* Philadelphia: P. Blakiston's Son and Co., 1899.

PRIOR, R. C. A. *On the Popular Names of Garden Plants.* London, 1879.

RHODE, ELEANOR SINCLAIR. *A Garden of Herbs.* Boston: Hale, Cushman and Flint, 1936.

RODALE, J. I. and staff, eds. *Rodale's Encyclopedia of Organic Gardening.* Emmaus, Pennsylvania: Rodale Books, Inc., 1971.

SIMMONS, ADELMA GRENIER. *Herb Gardening in 5 Seasons.* New York: D. Van Nostrand Company, Inc., 1964.

SMITH, A. W. *A Gardener's Book of Plant Names.* New York: Harper and Row, 1963.

SMITH, JOHN. *A Dictionary of Economic Plants.* London: Macmillan Co., 1882.

STEDMAN, THOMAS LATHROP. *A Practical Medical Dictionary.* Baltimore, Maryland: William Wood and Co., 1934.

TERRY, JACK R., publisher. *Herald of Health* (various issues). Mount Ayr, Iowa: Paragon Publications.

WARREN, IRA. *The Household Physician.* 1860.

WOOD, H. C., et al., eds. *The Dispensatory of the United States.* 22nd ed. Philadelphia, 1937.

WOODVILLE, WILLIAM. *Medical Botany.* Vol. 1. 1810.

YEMM, J. R., ed. *The Medical Herbalist.* National Association of Medical Herbalists of Great Britain, 1935–1937.

YOUNGKEN, HEBER W. *Textbook of Pharmacognasy.* Philadelphia: The Blakiston Company, 1948.

INDEX

Note: The following is an index of garden plant uses only, since the plants themselves are listed alphabetically in this book.